I0878592

The Forgotten Sense

The

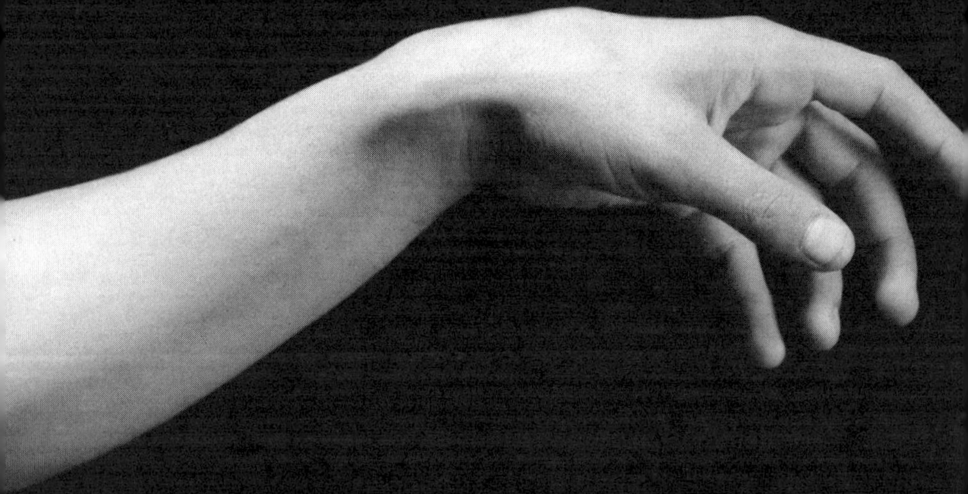

PABLO MAURETTE

Forgotten Sense

Meditations on Touch

THE
UNIVERSITY OF
CHICAGO PRESS
Chicago and London

The University of Chicago Press, Chicago 60637
The University of Chicago Press, Ltd., London

Published 2018
Printed in the United States of America

Spanish edition:

Translated, revised, and expanded by the author.

27 26 25 24 23 22 21 20 19 18 1 2 3 4 5

ISBN-13: 978-0-226-56133-2 (cloth)
ISBN-13: 978-0-226-56147-9 (paper)
ISBN-13: 978-0-226-56150-9 (e-book)
DOI: https://doi.org/10.7208/chicago/9780226561509.001.0001

Library of Congress Cataloging-in-Publication Data

Names: Maurette, Pablo, 1979– author.
Title: The forgotten sense : meditations on touch / Pablo Maurette.
Description: Chicago ; London : The University of Chicago Press, 2018. | Includes bibliographical references and index.
Identifiers: LCCN 2017038267 | ISBN 9780226561332 (cloth : alk. paper) | ISBN 9780226561479 (pbk. : alk. paper) | ISBN 9780226561509 (e-book)
Subjects: LCSH: Touch. | Senses and sensation.
Classification: LCC BF275 .M397 2018 | DDC 152.1/82—dc23
LC record available at https://lccn.loc.gov/2017038267

⊗ This paper meets the requirements of ANSI/NISO Z39.48-1992 (Permanence of Paper).

For Erin June Lodeesen

CONTENTS

Ich fühle mich!
Ich bin!
—Johann Herder[1]

PREFACE

Let's begin with the title, both nostalgic and misleading: *The Forgotten Sense.* But touch was never really forgotten. And it is also not *a* sense: it is many. Although it's undeniable that, from very early on, Western culture has been especially fond of light and sight, of eyes and clarity; although it is a fact that some of the most influential philosophical and aesthetic traditions of the past 2,500 years were particularly reticent about conceding any kind of virtue (intellectual, aesthetic, moral) to the sense of touch, and some even reviled it unceremoniously; although it is noteworthy that in the last fifteen to twenty years, intellectual historians and critics have written insistently about the oblivion of touch and have called once and again to repair such neglect: in spite of all of this, touch was never really forgotten.

Since the early days of Western culture in ancient Greece, a number of ways of understanding and representing reality that privileged the sense of touch, the body, and the material world has coexisted with that which some have liked to call *ocularcentrism,* a pervasive tendency to place sight over all the other senses as the most noble and trustworthy. These philosophical and artistic currents that underscore the role of tactility run parallel to the mainstream forces of ocularcentrism in the West, at times through subterranean channels, vilified and even persecuted, and at other times as dominating tendencies. However, and more important, touch was never really forgotten because remaining impervious to its ubiquitous effects is, quite simply, impossible. Plato can bypass it and, instead, praise sight as the sense that is most akin to the intellect, but when he describes the epiphanic moment when the soul encounters the divine, the language of tactility barges in uninvited. Christianity can deem it the dirtiest and most dangerous of the senses, the gateway to temptation, the instigator of unspeakable sins, but when it is time to account for the personal experience of true faith, the dogma of incarnation, the mystery of the cross, and the rejoicing in the divine, tactile metaphors and images find their way in unrepentantly to illustrate, with unrivaled perspicuity, the essence of religious fervor. This is because touch is not one sense; it is many.

Touch is the external, epidermal sensation of the outside world and also the intimate experience of our inner body. It is the sense of pleasure and pain in all their dizzying array of degrees and forms. It allows us to perceive the outside world not only as texture but also as pressure and temperature. It collaborates with the other senses to orient us in space and grants us the perception of our own bodies as living organisms. The vestibular sense by which we gain equilibrium, the sense of movement through space, the acceleration and deceleration of the body, and proprioception, the sense of our body parts in relation to one another and to their surrounding objects, are also variants of touch. Last but not least, touch is the sense that governs affect. Everything that moves, thrills, agitates, and inflames us, everything that causes in us even the slightest affective movement, is ultimately experienced as a form of touch. So even if we try to ignore it, touch, that many-headed hydra, elusive as a bar of soap under water, yet omnipresent like the very feeling of being alive, is a catalyst of existence and, as such, it is unavoidable, inexorable, and impossible to forget.

In recent years, staggering advances in technology have led to more and more voices that warn about the dawn of an age of detachment in which human beings will lose touch with one another and will be stripped of the ability to feel empathy, to achieve real intimacy, and to experience compassion. No one expressed this better than Leonard Cohen when, in reference to the protocols of prophylaxis made mandatory by the spread of HIV, he sang: "Everybody knows that the naked man and woman are just a shiny artifact of the past." But in spite of latex and pandemics, in spite of telecommunications, the increasing isolation in big cities, and the tyranny of the image, touch can be neither lost nor forgotten, nor overcome, nor avoided in any conceivable way. Just as we are not able to jump over our own shadow, we cannot lose the capacity to feel, to be affected, and to affect the world and those who live in it.

Given that when we talk about touch we are talking about a complex sensorial system, an aesthetic approach to the subject seems the most appropriate. The idea of a series of essays that explore the ways in which literature, philosophy, and art make use of, engage with, and evoke the senses of touch was born, in part, from a suspicious reading of the first chapter of *Mimesis* (1946). In it, Eric Auerbach draws his famous distinction between two types of narrative that constitute the backbone of Western literature: the Homeric poems and the Old Testament. Whereas the Homeric style (clear, exhaustive, detailed) betrays a necessity to externalize phenomena, thus making them accessible to our senses, that of the Hebrew patriarchs is the linguistic version of the chiaroscuro in painting: the darkness of the background and the opacity of the context are deliberate and aim at directing the attention toward the lesson that the passage hopes to teach. Auerbach's distinction is by all means stimulating and it produces exegetic results of enormous value, but it stems from a conception of the literary that is overwhelmingly visual. In Auerbach's view, the Homeric narrative transports us to a "realm where everything is visible,"[2] whereas the style of the Old Testament consists in games of lights and shadows akin to those later mastered in the visual arts by Caravaggio and Rembrandt.

Literature, film, painting, and sculpture, however, do not just appeal and cater to vision and do not just utilize narrative and stylistic techniques modeled on sight. The Homeric poems belong to ancient traditions of oral poetry that go back over three millennia. The revolution of linear perspective in the Renaissance is, in many ways, the expression of a tactile anxiety and a yearning for depth and three-dimensionality. And photography (along with its rich cousin, film) has a large variety of resources to reproduce and evoke the nuances of depth and texture. The essays that compose this book hover over the cultural, literary, and intellectual history of the West, lingering on moments that are especially revealing of the ways in which the many variants of what we call the sense of touch are engaged with, imitated, explored, or incorporated as formal elements. I am as interested in works that draw attention to the issue of touch, scrutinizing it and dissecting it, as in anything that unfolds in a tactile way—texts and works of art that express themselves by touching and pulsating, by beating, twisting, expanding, and accelerating, by warming up and cooling down, by gaining and by losing balance. Language and art are capable of all this and much more.

I began writing *The Forgotten Sense* as I was finishing a doctoral thesis in comparative literature on the revaluation of the sense of touch in the Renaissance. These essays became a refuge from the constrictions imposed by academic discourse; a refuge both methodological and stylistic. To a large extent, this book is the product of an urgency, of a creative and aesthetic need. For that reason, a caveat is appropriate. The wide range of sensations and emotions that literature and art produce happens in a liminal space, and the aesthetic effect manifests itself (it is *felt*) as an intermittent tickling, as an affective titillation that inspires and repels, that moves and shakes. Whoever writes about these matters runs the risk of falling prey to self-indulgence, of being carried away by streams of introspection until being irredeemably lost in labyrinths of digressions concerning the pleasure produced by aesthetic enjoyment. I have tried to avoid this. But because the borderline space between the work of art and its recipient, like all zones of frontier, is both fascinating and revealing, the essays that follow engage with a selection of literary and philosophical texts, historical phenomena, and cultural landmarks, at times brushing slightly against them, at times digging into them, or pressing on them, or connecting them with others that may at first appear unrelated. One of the goals of this book is to call attention to (pardon my French) *la texture du texte*, the skillful embroidery of formal elements and content that creates the unfathomable universe of a text. The only way to gain access to such textures is by touching them and by being touched by them. Should these essays manage to identify narrative and stylistic techniques, transhistoric trends, and poetic forms of a tactile nature, they will have, perhaps, made a humble contribution to the study of aesthetics. But if, even for an instant, they succeed in expanding until they stroke the erogenous frontier where the magical encounter between language and those sensations that transcend it takes place (what I mean, reader, is if at any given moment my words manage to touch you), then, and then only, will they have accomplished something that for lack of a better word I dare call *true*. Without further ado, let this be a welcoming squeeze of the hand.

Chicago, May 30, 2017

A SQUEEZE OF THE HAND

The man has no eyes. He arrives at the brothel, and upon being presented with the selection of choices he proceeds to feel his way around them. He touches their limbs and strokes their skin, brushes their hips, grabs, pokes, and decides which one is the softest, the whitest, the most beautiful. By touch he discerns, touching he assesses. Eventually, he picks one; his *docta libido* has endowed him with many eyes, why would he want two more? Luxorius, the forgotten Latin poet, poses this question in one of his epigrams. In another poem, he praises a famous gladiator of his time, whose dexterity with the spear and the sword was such that, in a celebratory portrait, the artist had depicted him with eyes on the palms of his hands. One might think that these two scenes reflect the poet's agreement with the traditional view according to which sight is the most precious and trustworthy among the senses. The blind man has more than two eyes, the gladiator sees with his hands. However, when Luxorius uses eye imagery, he almost seems to be contesting, even parodying, the even then already long-standing belief in the preeminence of sight. The blind man's tactile perception, certainly hyperdeveloped to compensate for the visual handicap, endows him with a sensibility that is richer, more nuanced, and more complex than that of someone who sees with merely two eyes. The gladiator's infallible technique resides in his hands, in his movements, in his posture and balance. Rather than celebrating eyes, the poet is stressing the limitations of the visual.

Luxorius lived in Carthage in the fourth century AD, when the city was part of the Vandal empire. His work—lost for over a millennium—recreates with striking vivacity this extraordinary world saddled between antiquity and the Middle Ages. A journey through his ninety-one extant epigrams is a fascinating promenade down the streets of Vandal Carthage, a city that Luxorius portrays with realism, humor, and compassion. Characters like Lucius, the obese falconer, Zenobius, the bad poet, Syracus, the gambler, Gattula, the clumsy ballerina, and Marina, the unfaithful wife, materialize and come back to life in his verses. The men and women, the statues, the frescoes, the houses of that Carthage inexorably lost in time, rebel against the forces of oblivion and resist in all their tangible reality. Luxorius stands out for his ability to recreate the very textures that make the fabric of everyday life. But this is not what makes him a rarity. After all, his poetry is part of a long tradition that goes back all the way to ancient comedy and is continued in classical Latin poets like Catullus, Ovid, and Martial. If I bring up Luxorius here it is because of his curious habit to reflect upon the relationship between the senses, and because of his appreciation of the importance, complexity, and self-sufficiency of touch.

Let's return to epigram 71:

In need of some light, losing
His way, the blind, uncertain
Lover with the widowed face gently
Touches and strokes the skin
And examines the limbs
Of the women, to judge, for himself,
Which are the most
Beautiful, which are snow white.
Skillful lust [*docta libido*] has given him

So many eyes, why should he want
Two more, simply to see?[1]

The tactile sensibility of this blind whoremonger reveals a number of things. First, sight has been overappreciated. In the beginning we are told that he is "in need of" (*egenus*) light and that his face is "widowed," or "void" (*viduae frontis*), which presumably means that his eyes have been gouged out; but this simply describes his lack of sight. Soon we will find out that this man not only does not need it but, were it up to him, he would rather remain blind. And the reason for this is that, second, touch is not one sense, but many. It enables the perception of textures, temperatures, silhouettes, forms, and even colors, thus allowing the blind man to composite a mental perception of tactile nature that helps him consolidate the type of aesthetic judgment needed to make a rational decision. But the most striking suggestion here is that this array of faculties that compose the sense of touch, and that assist the blind man as he decides who his companion will be, has its origins in a "skillful lust" (*docta libido*). *Libido* in Latin, as well as in English, refers specifically to sexual desire, but also to will and yearning in general. It is a term that brings together the spheres of affect and volition. In fact, the blind man's libido is not ignited when he begins stroking and feeling bodies. The man arrives to the brothel moved by this lust that—one might presume—took possession of him suddenly and unexpectedly, as often happens with desire. Furthermore this libido is a *docta libido*—an astute, expert, trained, skillful kind of lust. But isn't this an absurdity? Isn't lust by nature irrational, flighty, unhinged? Not in this case. This kind of *libido* is all too familiar to the blind man; he has experienced it before. This *libido* is of a restrained kind, the man knows how to steer it and where to direct it. It is a *docta libido*, a learned lust. Just like touch can be trained and refined until it becomes a sense so subtle and comprehensive that it supplants sight, so too can lust be educated. Through his portrait of the blind man at the brothel, Luxorius suggests that behind, or underneath, touch understood as mere physical contact, there is something more primordial that functions as a principle of movement, an affective engine that can be operated in manners that vary in their degree of rationality, with more or less skill.

Since "touch" is mainly associated with physical contact on a superficial, epidermal level and, in particular, with the sense whose organ is the skin, the word proves insufficient when it comes to affect, the sensation of one's own body, as well as in reference to other faculties alluded to in Luxorius's epigram. And in spite of the fact that language itself, understood as the millenary sediment of fossilized metaphors and forgotten meanings, reveals the astonishing polysemy of the word "touch," this conceptual deficiency is crippling. The word "touch" comes into English through the Old French verb *tochier*, from the Latin *tactus*, participle of the verb *tangere*, and it has the longest entry in the *Oxford English Dictionary*. From very early on, apart from its literal significance, it meant "to pertain to," a type of contact without actual contact, a connection based on affinity. So too, when a word, a melody, a gesture, an image "touches" us, we are referring to some form of contact without actual contact. Another even more interesting metaphorical sense of the word is that inherent in the notion of "tact." To act with tact is to be skilled in successfully having a certain influence on and in directing the emotions of others. To act with tact is to know how to manipulate affect; it is a form of touch that dispenses with physical contact. Eloquent and rich though these senses of the word and its derivations may be, they are still metaphorical. And although a large number of terms in any given tongue are archaic metaphors, whose literal meaning has been forgotten in the course of the many and constant semantic mutations that languages go through as they evolve, in some, as in "tact," or in "touch" understood as affect, the metaphoric element is still prominent and dominating. In order to explore the range of varieties of the tactile phenomenon, it is preferable to employ another concept, that of the *haptic*, whose relative novelty ensures a level of literality that helps to avoid misunderstandings and ambiguities. Thanks to the strangeness that its newness and artificiality arouse, this concept creates the distancing needed to ponder something as familiar and immediate as the question of touch from a different angle. The notion of the haptic possesses three great advantages: first, it is unfamiliar and contrived enough to create estrangement and demand attention. Secondly, its relative novelty guarantees (for now) a certain degree of stability and impermeability vis-à-vis the inexorable semantic mutations that characterize the history of every word. Finally, it is an elastic enough concept that, in the course of its short history, has expanded rather than shrunken its umbrella of meanings. The notion of the haptic includes touch understood in its literal and metaphorical sense and cannot be but thought alongside the tactile, but its artificiality, its impermeability, and its elasticity make of it an invaluable navigational instrument for whomever sets out to explore the ocean of tactility, of which skin is only the surface.

This tendency to nuance and expand the notion of "touch" is not unique to our era of distraction and multimedia. The development of virtual reality, experiments in teletactility, three-dimensional images and printing, haptic interfaces, and other such twenty-first-century extravagances are the expression of intellectual and aesthetic concerns as old as human beings themselves. The visual arts have always looked for tricks to imitate depth and three-dimensionality. From late Roman bas-relief sculpture to Anish Kapoor's monumental structures and installations, from Zeuxis's grapes to the (re)discovery of linear perspective in the fifteenth century and to the invention of film and holography in the nineteenth and twentieth centuries, these are all examples of how plastic arts respond to the same demiurgic impulse—the impulse that moved Pygmalion and Victor Frankenstein—by which we strive to create artificial representations of life that are as realistic and tangible as possible. In the last couple of decades, the development of technology that reproduces tactile sensation has skyrocketed in the form of ever more sophisticated and subtle touchscreens and tactile interfaces, computer games and watches that allow us to share our heartbeats with others, sex toys that work remotely, forms of media that allow for teletactility, and even prosthetic limbs that restore sensation. So too, we have seen the increasing engagement of art with senses other than sight, and especially touch in growing numbers of artistic exhibits that propose a tactile involvement with the visitor, be it by touching the works of art themselves or by entering them and experiencing their space and atmosphere. Needless to say, scholars and intellectuals have echoed this haptic enthusiasm. From fields as far apart as art history and medicine, philosophy and history of science, anthropology, psychology, cognitive theory, and others, more and more research has been mounting on the topic of the senses in general, but, more specifically, on the topic of the senses that were historically neglected, like touch and smell.[2] It is in this context that the notion of the haptic proves to be a fruitful theoretical framework from which to consider the sense of touch in innovative ways, while still accounting for its multifarious nature.

In 2005, historian Robert Jütte announced that the twenty-first century marked the beginning of a "haptic age."[3] Over a decade has passed since the bombastic announcement and the interest around the haptic has not given any signs of declining. The term "haptic" comes from the Greek verb *haptomai*, which means to "come into contact with," "to touch," "to grab." The deponent character of the verb (active in meaning, but bearing the form of a verb in passive or middle voice) reflects one of the most fundamental features of the haptic: its simultaneity.

To touch is to be touched. To feel something is also to feel oneself. The notion of the haptic results from a growing suspicion among intellectuals that touch is exceptional among the senses. Its reluctance to schematization, its fleeting and versatile nature, and its peculiar relation with the other senses were the first aspects that drew the attention of those who pioneered the positions that would eventually lead to the coining of the term "haptic." In his *Essay Towards a New Theory of Vision* (1709), George Berkeley elaborates on an intuition that Descartes before him had introduced in his own work on optics: touch and sight collaborate. According to this idea, which one can actually already find in classical and medieval works on vision, the senses are not independent and discrete faculties, they converge, they overlap, they assist, and they compete with one another. A few decades after Berkeley, Johann Herder, in his treatise on sculpture (*Plastik*, 1778), went deeper into the notion of the complexity of the tactile experience and placed the focus of his aesthetic theory on the relationship between sight and touch. Finally, toward the beginning of the twentieth century, art historian Aloïs Riegl introduced the term "haptic." After a long career as curator of textiles at the Imperial Royal Austrian Museum of Art and Industry, Riegl developed a theory of aesthetics that conjugates formalism and historicism based on the observation of textures and surfaces to detect stylistic patterns across cultures through different historical periods. According to Riegl, the haptic is related to two characteristics of the work of art and the cultural context in which it is produced: the conception of space and the distance that must be kept between the spectator and the artifact. The more disconnected from its surrounding space an artifact is, the more haptic-sensitive the culture that produced it proves to be. By blurring the connections with and reducing the importance of the spatial context, attention is directed solely to the object itself in its unique texture and design pattern. In order to access the aesthetic universe of such artifacts, the spectator must approach them, bracketing as they do so the surrounding space.[4] Riegl also talks about "haptic devices," such as overlapping and foreshortening, that by appealing to a type of sight from up close that caters to tactility "create the sense of spatial recession."[5] On the other hand, when the artist strives to create a spatial coherence within the work, and situates the different elements as organic parts of a whole, the distance required to understand the work increases. This tendency Riegl ascribes to what he calls optic cultures. In the triumph of linear perspective and other techniques of trompe-l'oeil, as well as in the tendency toward abstraction, Riegl finds confirmation of the preference for optic vision in Western art. Some of Riegl's contemporaries, art historians like Wilhelm Worringer and Bernard Berenson, also explored the notion of a tactile vision in their studies of painting and sculpture, the latter focusing on the Italian Renaissance, but after them this approach and even the term "haptic" are mostly relegated to oblivion. In the second half of the twentieth century, Gilles Deleuze goes back to Riegl and borrows the notion of the haptic to analyze film and modern painting. Deleuze believed that the greatest advantage of the term resided in its ability to overcome the dichotomy sight-touch. Some of Robert Bresson's films and some works by Francis Bacon construct, says Deleuze, a space in which only haptic vision can orient us—one that demands from us a type of seeing that is like touching. In *A Thousand Plateaus* (1980), Deleuze and Guattari also borrow Riegl's notion of the haptic as they introduce the distinction between the smooth and the striated, two concepts that account for close and distant vision. Paradoxically, Deleuze overcomes the duality sight-touch by introducing a new duality: optic and haptic vision; and the haptic engenders a new form of duality (rizomatic, but dual nonetheless): smooth-striated.[6]

The notion of the haptic was introduced to account for a certain type of synesthesia between the senses of sight and touch by means of which we are able to anticipate the texture of things without having to touch them. We do this thanks to a repository of data that we collect from very early on in our lives. We don't need to touch sandpaper to know how rough it feels, or velvet to know how soft and smooth it is. We know these things thanks to our haptic memory. In the past decade, however, the notion of the haptic has expanded to include other forms of sensitivity that transcend the sphere of epidermal contact (*exteroception*), as well as the space where the negotiations between sight and touch take place. *Intraception* (the perception of the interior of the body), *proprioception* (the perception of the different parts of the body in spatial relation to one another), and *kinesthesia* (the perception of the movement of the body) are also considered haptic faculties. It is hard to deny that swallowing and vomiting, chewing, eating, defecating and excreting in general, and sensations such as a stomach cramp, a pang of headache, the beating of the heart (any form of internal pain and pleasure in general, for that matter) manifest themselves in tactile ways. And, elusive, ungraspable even, though the feelings of balance, bodily orientation, movement, acceleration, and deceleration of the body are, they too produce a type of sensorial feedback that one can associate with the tactile, rather than with any other of the senses. Ultimately, all these forms of bodily perception are associated with the sense of touch and fall under the notion of the haptic because their nature is, first and foremost, affective.

Affectus, from the verb *afficio* (*ad* + *facio*), refers to any action that produces a change of state, any action that leaves a mark, that perturbs, that initiates movement, and that touches. Affect is not merely another variant of the haptic: it is its touchstone and its ground zero. Affect is movement, but it is also the cause of movement. Affect is the position of each part of the body at any given moment in regard to the other parts, and also the outside world: it is our most primordial form of proprioception, and it constitutes the unshakable center of the compass that orients us in the world. Affect encompasses every form of external and internal sensitivity, but it also refers to that which precedes and makes sensitivity possible. Affect is the mysterious core of life, a spontaneous mechanism, an unmoved mover that comes into action with the spark of conception and shuts down right after the hosts of death have taken possession of the body. To be and to perceive oneself as a living being, as an animated body inhabiting the world, is the most basic form of affect. Some of the most enlightening accounts of affectivity in the twentieth century are Heidegger's notion of *Befindlichkeit* and his analysis of moods and feelings, Merleau-Ponty's inquiries around one's own body (*le corps propre*), and Michel Henry's phenomenology of the flesh, especially his idea of self-affection (*auto-affection*), a foundational experience that reveals the very fact that we are alive, and that lies at the basis of all subsequent sensitivity. More recently, in the Anglo-Saxon world, the proponents of a so-called "affect theory" (Brian Massumi, Silvan Tomkins, Eve Sedgwick, Lauren Berlant, and Eugenie Brinkema, to name just a few) have explored the moods overlooked, or merely mentioned, by phenomenology in literature, in art, in film, in culture and society. Philosophers and theorists, however, tremble before the prospective of not getting to the bottom of things. The fear of taking complex entities for simple ones, and being accused of myopia or, worse, of naiveté, often paralyzes them. For this reason, philosophy and theory feel more at ease among forms and paradigms, in a world of a priori knowledge and transcendental categories, a world beyond sensation and particularity, a world seen long distance, a world put into perspective and conceived panoramically from theoretical vantage points. It is not surprising that the language of theory and philosophy is predominantly visual. But the world is a never-ending carnival of the particular, the uneven, and the irregular—an infinite bazaar of unique textures that cannot be duplicated, an uncharted continent that only reveals its secrets to those who inhabit it by groping about in the dark. The notion of the haptic offers a way that leads from the proto-sensitive depths of our affective core, through intermediary stages (that are sensible, yet virtually imperceptible) such as proprioception, to the myriad tangible manifestations of affect and of sensibility in general.

Its conceptual plasticity and its ability to account for the nature of affect are perhaps the two most noteworthy advantages of the concept of the haptic. Also, if we understand it as a conglomerate of faculties, both sensorial and constitutive of the very nature of sensitivity, it becomes clear that the haptic is exceptional among the senses. Five basic reasons for this come to mind. To begin with, it is the first sense (or, rather, sensory collective) that is activated in the quasi-mythical moment when the embryo first feels itself as a living entity. Exteroception is inaugurated around the eighth week of gestation when the skin, the organ of touch, starts forming as an epidermis consisting of basal layer and periderm. The haptic is, therefore, the cornerstone of human experience in the world. Second, it is the only sense that one cannot lose. By merely being alive we are feeling and we recognize ourselves as living bodies in the world. Even in cases of deep coma there is enough brain activity to allow for certain forms of sensation and reaction—it is not uncommon for doctors to encourage people to touch and talk to their comatose family members. Third, the haptic is the only form of sensation that is not localized in one specific part or organ of the body. It originates in our fathomless interiority and operates in every muscle, organ, and ligament; it runs through our bloodstream following the rhythm of our heart, and it spreads throughout our skin (the vastest, most complex organ in the human body), taking roots in our affective core. Fourth, it is the only sense that can unfold itself. When we feel our inner body and when we perceive the movement of our limbs, there is a coincidence between that which feels and that which is felt. When we touch ourselves, we are both touching and being touched. A simple experiment like scratching our right hand with our left puts into question basic categories such as subject and object, active and passive, opening new conceptual possibilities and suggesting that there are alternative ways to understand corporeality, subjectivity, and identity. Finally, the haptic does not need a medium, such as air or light. Instead, it operates in the most absolute immediacy. Its affective nature is strictly connected to this immediacy, which, in many ways, constitutes the last line of defense between the outside world and our interiority. When an image, a melody, or an aroma, moves, frightens, disgusts, or upsets us, the feedback is invariably haptic: we get sick to our stomach, our heart races, our temples pulsate, we get goose bumps, we feel pressure in our chest and a commotion in our lower abdomen, we shake, we heat up or cool down, we sweat, we tear up, we choke.

The notion that touch was exceptional among the senses had already a long history when Luxorius wrote his epigrams in Carthage among the Vandals. Aristotle, the first to introduce the distinction between the five senses, wondered whether touch was one sense, or many. Fond as he was of symmetries and dichotomies, the irregularities of touch made him uncomfortable. If the organ of sight is the eye, and its medium, light, what is the organ of touch? And its medium? The flesh, the skin maybe? If sight oscillates between two extreme opposites, black and white, hearing between acute and grave, and tasting between bitter and sweet, what two extremes mark the spectrum of touch? Aristotle concludes that it must surely be a special sense disseminated throughout the body.[7] As the sense that we share with all animals, touch is also the most basic and primitive sense, and by no means the most noble, the philosopher adds. When he analyzes the senses in his treatise *On the Soul,* he begins with sight and ends with touch, thus establishing a hierarchy that would live on well into early modernity, and beyond. The famous opening lines of the *Metaphysics* confirm this belief in the superiority of sight over all the other senses.

There Aristotle claims that, of all the senses, sight is preferred by human beings because it is the one that allows us to know in the clearest and most exhaustive way; sight allows us to draw distinctions, separate, categorize. And in the *Nicomachean Ethics*, where the nature of the senses is broached in relation to the issue of pleasure, the philosopher repeatedly warns against the dangers of tactile enjoyment, the most pernicious and distracting of all. In spite of this, Aristotle also compares the soul with the hand since they both are an "instrument of instruments" (*organon pro organon*), associates manual dexterity in humans with intelligence and the superiority of the species, and, in the *History of Animals*, affirms that human touch is "the most accurate" (*akribestate*) of all the animal kingdom.[8]

Plato, the other great champion of ocularcentrism, also had conflicting positions regarding the sense of touch. He believed that sight was more akin to the intellect than any other sense and, therefore, the most conducive to initiating the process of leaving corporeality behind and finding refuge with the soul in the intelligible sphere. In the *Republic*, the sun and the light stand for truth and intelligence; the soul is freed from its bodily chains when the prisoner leaves the cave and sees the light of day. The *Timaeus* includes a passionate praise of sight, source of all our knowledge about the universe and springboard to a succession of stages of increasing metaphysical awareness. As Plato discusses the generation of the material world and of human beings, and as he tackles the issue of the senses, touch is not even mentioned. In the *Symposium*, when Socrates arrives at the gathering and Agathon, the host, invites him to sit by his side so that, by contact with him, he can absorb some of his wisdom, the philosopher replies: "How splendid it would be, Agathon, if wisdom was the sort of thing that could overflow from the fuller to the emptier of us when we touch each other."[9] And yet, at the most mystical moment of the dialogue—and, perhaps, of all of Plato's work—when Diotima describes the instant of communion between the soul and the form of beauty, she says that it is a "wondrous vision" that "one is almost able to lay hold of" (*schedon an ti haptoito*).[10] If the soul and the divine are to become one, there can be neither distance between them nor any kind of duality left, two necessary conditions of all seeing. The light of the divine is blinding, and the pilgrim soul must grope about like the blind man in the brothel. Thus, it is not surprising that, when it is time to describe that which is indescribable, the overcoming of all duality and the breaching of all distance, Plato appeals to the language of tactility. Plotinus, his most brilliant successor and father of rational mysticism in the West, follows in the master's footsteps. "Anyone who has seen it knows what I mean when I say that it is beautiful," he says regarding the One, divine origin of all being, life, and thought; and, when he deals with sense perception, he discusses sight, hearing, smell, and taste, but leaves touch unmentioned.[11] As for Plato before him, the progress of ascent toward the realm of the divine takes place by means of vision, first sensible, then intelligible. This notwithstanding, when he attempts to account for the moment of communion between the soul and the One—a task that is fundamentally impossible, given that through such union all duality, and, therefore, language itself, is overcome and annulled—Plotinus says that the One makes itself available only to those who are capable of "something similar to grasping it" (*hoion ephapsasthai*) and, later, "touching it" (*thigein*).[12]

Thus, for Plato the highest possible epiphany is an "almost" touching, whereas for Plotinus it is "like" touching and grabbing. The irony is abysmal. By means of metaphors and idioms, the sense that they both most deplored sneaks into their discourse to account for the experience that marks the culmination of the philosophical endeavor. The reason for this is that touch, which is not one but multiple senses, is said and meant in many ways. Although Aristotle realized this early on, no one in antiquity dug deeper into the matter than the atomists. According to Diogenes Laertius, Epicurus had written a treatise *On Touch* (*Peri haphes*), that has not survived; and in *On Sense and the Sensible*, Aristotle criticizes Democritus and atomists in general for claiming that all senses are variants of the sense of touch.[13] A few centuries later, in Rome, Lucretius's philosophical poem *On the Nature of Things* (*De rerum natura*) provides the most exhaustive treatment of the primacy of touch in all antiquity. For Lucretius touch (*tactus*) refers to, at least, four faculties. First, *tactus* is the activity of constant collision in which atoms engage through the void. Since atoms are unchangeable and insensitive bodies, this form of touch is characterized by a failure to affect them and by a failure to produce any changes in them. It does, however, bring them together and the result of this perennial gathering is the formation of composite bodies. Second, *tactus* refers to physical contact between composite bodies. This touch does affect and it is through it that bodies reproduce, multiply, wear out, and, eventually, come undone as their atoms move on to form other bodies. Third, *tactus* means sensitivity in general (what Lucretius calls *sensus*) because, for Lucretius, any form of feeling is a form of touch. Finally, it is one of the five senses whose organ is the hand.

Lucretius believed that *tactus* rules the nature of things. The universe is divided into atoms and void; that is, bodies and the space in between them that allows them to move around, come together, and disaggregate. Touch is at the bottom of the very notion of body and materiality. Lucretius cannot be any clearer: "For nothing can touch or be touched except body."[14] Void, on the other hand, is essentially intangible, or untouched (*intactus*): "A property is that which without destructive dissolution can never be separated and disjoined, as weight is to stone, heat to fire, liquidity to water, touch to all bodies, intangibility to void."[15] And given that generation and corruption is a tactile affair, so too is knowledge. All senses are variants of touch. Vision operates by means of images that come off of things, like bark on trees, and impact against the eye. Taste functions thanks to minuscule bodies that touch the tongue; smell, to odorous atoms that penetrate the nostrils; and hearing, to atoms of sound that enter through the ears, resonate in its cavernous geography, and collide against its walls. In being each and every one of the senses, touch is sensitivity itself and this realization leads Lucretius to a rare outburst: "For touch, so help me the holy power of the gods, it is touch that is the bodily sense!"[16]

In the atomistic universe everything is at once tactile and tangible. Atoms, insensible as they are, touch each other. Sensible bodies touch each other. Thinking and knowing are tactile faculties as well; also dreaming, fantasizing, philosophizing, and poetizing.[17] Movement is also tactile. That mysterious force that opens our eyes in the morning is touch. The impulse that makes us get up from bed is touch. So too, the maritime tides, the flight of birds, the orbit of planets and celestial bodies. Touch both precedes and transcends sensitivity, constituting it. It is the cause of all movement, it rules the generation and corruption of composite bodies, it is the catalyst for the collisions and encounters of atoms through the void, and it is what enables physical contact between the outside world and us. The Lucretian notion of *tactus* is thus the clearest, most direct predecessor of that of the haptic. Despite the complexity of the tactile sensorium (or, rather, *because* of it), it is not surprising that atomism is one of the few philosophical traditions to have emphasized the importance of tactility. To think about touch and the haptic is a daunting task if we consider that Western philosophy is, to a large extent, modeled on vision. To theorize (from *theoreuein*, "to contemplate") is to see with the intellect; an idea (from *idein*, "to see") is something seen with the eye of the mind; to categorize, to draw distinctions, to taxonomize are processes that rely on clear separations between parts, categories, subject and object—processes that necessitate distance and perspective. Is it at all possible to philosophize blindly?

Can we draw distinctions and categorize from up close? Can we think *haptically*?

Lucretius certainly believed it was possible. To him, the operations that allow for thought are tactile, and ideas, like everything else in the universe that is not void, are tangible material realities. A term like "concept" (from the Latin *concipio* and *capio*, "to hold," "to grasp") and its derivatives highlight that metaphorical affinity between understanding and the use of the hand that Aristotle had already pointed to.[18] To conceive, in this sense, has a purely practical significance, because it does not mean simply to capture something, it means to hold on to it and use it appropriately, dexterously. The haptic must then be a matter for practical, not theoretical knowledge; "haptic theory" is, therefore, an oxymoron. However, any form of practical knowledge is also a form of specific knowledge, knowledge about a singular practice (masonry, bakery, saddlery, poetry, pedagogy, etc.) acquired over a considerable period of sustained engagement with it. Is a science of the haptic in general possible, then? Is it conceivable to imagine a type of knowledge that, without neglecting the particular, manages to account for the universality of the haptic? Most likely, no. Striving to apprehend the haptic by means of thought and language might just be like trying to grab a bar of soap under water. Perhaps the matter is simply not suited for operations like thinking and knowing, but rather for sensing, and the haptic not a problem for philosophy, but one for aesthetics. After all, it was an art historian who coined the term.

From the morning of modernity comes an emblematic intimation of haptic aesthetics. It involves a man fondling a hermaphrodite. The scene takes place in Rome, at the Basilica of Santa Cecilia in Trastevere. The year is 1440 and the man is sculptor, goldsmith, architect, and art critic Lorenzo Ghiberti. His specialty is bas-relief and his claim to fame, the biblical scenes on the doors of the Baptistery in Florence. The hermaphrodite is a second-century-AD Roman statue, which had been recently unearthed from an ancient sewer. It is the size of a prepubescent boy, or girl, and it has been carved with such mastery that, upon seeing it, Ghiberti remains in awe. The androgynous figure is covered by over a thousand years of dirt that some servant or other carefully wipes clean in the courtyard of the basilica. Once the statue is free from stains, Ghiberti approaches it and travels its marble contour with his hands. Only then he understands that it is a hermaphrodite. Years later, in his *Commentaries*, he writes: "It is not possible to express in words the perfection of this statue . . . which, covered in a subtle cloth, showed the male and the female nature. . . . Many pleasant attributes did the statue possess, and none of them could be grasped by sight, unless the hand found it through touch."[19]

The hand brushes the cold and smooth texture of the marble, tarrying in its ambiguous curves and bulges, and guides the eye, which accompanies it, lagging behind. Together, hand and eye compose a mental likeness of the artifact that is at once visual, tactile, and spatial: a haptic representation. As he detects the genital peculiarity, Ghiberti grasps the essence of the work. Sight alone, deficient and misleading, would never have revealed it. Without saying it, perhaps without even thinking about it, Ghiberti was inaugurating haptic aesthetics. Toward the mid-sixteenth century, when painters and sculptors argue over whether painting or sculpture had the better claim to being superior among the plastic arts, the proponents of the latter (among them Benvenuto Cellini and Michelangelo) point out precisely this: touch does not deceive, sight does; touch gives us reality as it is, sight deforms and transforms it. Later on, Riegl would say that only the sense of touch allows us to apprehend reality as it really is, in three dimensions.[20] Among the arts, sculpture is the only one able to fully exploit the three-dimensionality of the world. It is therefore natural that, when faced with a statue one should be tempted to behave like Mozart's reckless Casanova in front of the *Commen-*

datore and reach out to touch it. And in doing so, perhaps we are obliging the artifact, whose extended hand and whose inviting smoothness often stimulate and incite tactility.

Many years before reading about Ghiberti's haunting encounter with the hermaphrodite, a very similar image had already inspired me to write about the sense of touch. It was a scene (or rather, a moment) from *Russian Ark*, Alexander Sokurov's 2002 tour de force (see fig. 1). In the course of the ninety-six-minute-long unedited shot, an anonymous and incorporeal narrator wanders through the Hermitage Museum by means of an ever-moving steadicam, an eye that never blinks, leading the spectator through three hundred years of Russian history and some of the most remarkable pieces in the collection. Another extemporaneous, yet visible, character accompanies the invisible narrator in his visit: the Marquis de Custine. Upon coming upon Antonio Canova's *Three Graces* (1817), the marquis and the camera slow down and circumnavigate the statue, mesmerized—the movement is hypnotic. They subsequently approach the marble, go up and down the silhouette of each of the three Graces and continue to orbit around them as if attempting to exhaust every single angle and viewpoint. Eventually, the cyclopean lidless eye draws even closer and it is at this point that the visual becomes tactile, as the camera subtly brushes against the sculpture in an almost ceremonial way. Finally, once satisfied, they both continue with their journey through the halls and rooms of the Winter Palace. A few minutes later, we see in the distance a woman touching the statue of a winged girl—it is Gennaro Cali's *Psyche Abandoned* (1832). The woman is wearing a mourning dress and it quickly becomes apparent that she is blind. She travels up and down the breast, face, legs, and arms of the statue with the tips of her fingers; her expression is of serene concentration and bliss ("I'm admiring the works of the great masters," she says to the marquis; he says to her: "Your skinny fingers observe everything"). As she conceives, she enjoys; as she enjoys, she con-

1. *Russian Ark* (dir. Alexander Sokurov, Russia, 2002).

ceives. The woman's roving hands illustrate, exemplify, or clarify what the camera had just done with Canova's rendition of the three daughters of Zeus. Throughout the film, Sokurov repeats the technique of the haptic glance, by approaching and going around certain paintings, pieces of furniture, garments, and even people.

It goes without saying that touch completes the aesthetic experience of a sculpture. Who hasn't, at some point, suspected that statues have a life of their own? This is because statues evoke the very condition of being alive, they imitate life by imposing before us their dense and heavy existence, and by inhabiting a space that is effectively three-dimensional. Just as sculpture is the most tactile among the arts, painting and photography would appear to be the most visual. Music is patrimony of hearing, film of both (or so it seems). Literature, however, is different. Literary language is forced to traverse through the filter of eyes and ears (or fingers!). But, when it comes to the literary, sight and hearing are not much more than that: filters, vehicles, means, rather than the actual space where the aesthetic experience takes place. It would seem that, in order to produce an aesthetic effect, literary language ought to transcend the senses and fall into that complex lake system formed by fantasy, imagination, memory, reason, intellect, and other synaptic faculties. If it were so, literature would be the least sensorial of all the arts. Clearly, this is not the case, and the reason for it is that, as in all other art forms, in literature, aesthetic affection takes the form of a haptic occurrence. In the same way that certain images and melodies, after entering the body through eyes and ears, affect us deeply, creating inner commotion, certain pages, certain stanzas, certain words trigger reactions that it would be absurd not to identify with the realm of the affective and, therefore, with the haptic. If the peak of the aesthetic effect is the sensation of being touched, moved, and transformed by a work of art, all of the exteroceptive senses must work in tandem for this to take place. At the same time, the highest aspiration of any work of art is to come to life in an ineffable moment of intimate connection with its interlocutor—the story of Pygmalion is the mythologem at the core of Western aesthetics. For an aesthetic artifact, coming to life means reclaiming for itself a tangible space where it can touch and be touched, and where it can assert its transformative vigor.

Just as Sokurov manages to caress the statue with the camera, we can be touched by a simile or a rhyming couplet, by a metaphor, by the right adjective in the right place, by the perfect verb. We are touched in the distance; it is touch through affect, it is haptic. Literature can achieve this because the haptic manifests itself in language and through it, in the inextricable skein of content and form that constitutes the literary work. In short, literary language can evoke touch and appeal to it, because it is an affective expression of the human spirit. But this doesn't happen exclusively when literature thematizes touch, when it describes textures and sensations, and when it exploits the literary possibilities of the body; literary language can go beyond content and actually imitate haptic mechanisms in its form. A haptic style organizes space in idiosyncratic ways by means of rhetorical figures such as metaphor, simile, and metonymy, by reducing distance, by zooming into details and blurring contexts and backgrounds. It can emulate movement through changes in rhythm, crescendos, and diminuendos. And it can create a proprioception of the text articulated by structural symmetries and asymmetries. A haptic style turns the work into a tangible and tactile object, and the text, as Roland Barthes would put it, into a textile. Additionally, a haptic style invites a critical approach akin to it. In the preface to her 2002 collection of essays on film and video, Laura Marks celebrates the dawn of the age of haptic criticism.[21] Thanks to the work of Riegl, Berenson, Deleuze, Marks, and others, painting and film have been the objects of such haptic criticism. Not so much literature.

A haptic literary criticism should account for the vast array of manifestations of tactility, tangibility, affect, movement, etcetera, in literature, while both transcending the strictly dermic stage of mere contact (i.e., literality) and avoiding getting lost in the region of the purely metaphoric; it should allow the critic to create distance without interrupting contact. Haptic criticism should be an invitation to intimacy, a friendship with benefits between the text and the critic, an explosive collision between two surfaces, whose shock waves reveal unsuspected depths, unfamiliar sensations, and primordial affections. Although literature is born as poetry, novelistic prose is a particularly propitious space to attempt this critical approach. The modern novel is born in the context of the industrial revolution on the immediate bases of eighteenth- and early nineteenth-century sentimentalism and romanticism. From its early modern beginnings (*pace* Heliodorus) until the 1922 publication of *Ulysses*—arguably, the work that exhausts the genre by pushing it to the utmost limits of its possibilities—the novel combines a construction of vast, transhistorical universes with the excruciating exploitation of character psychology. The many branches of human knowledge that were, by the early nineteenth century, well on their inexorable way to almost complete atomization come together in its titanic embrace. In the brave new world of mass production and mass consumption, the novel rises as a stage where the phenomenon of proliferation and acquisition of objects and goods is exposed, glorified, criticized, and parodied, but, especially, registered. As the product of a world in which things were being multiplied *ad infinitum,* the novel indulges in detail, abounds in minutiae, and is often structured around meticulous descriptions of people and things. It portrays rings and necklaces, shoes, coats, gloves, hats, shirts, and undershirts; it goes past underwear and gets to the naked body, it conquers its hidden geography and accounts for every dimple and every sunspot, every freckle and every wrinkle, every stretchmark and every protuberance; and then it goes beyond the body and, thanks to one of its most staggering inventions—free indirect speech—arrives at the soul, psyche, mind, consciousness, unconscious, heart, subjectivity, affect, or wherever it is that all inner life is born. One of the most astonishing effects of this is that it positions people and things, feelings and actions, on the same plain—all of Charles Bovary is contained in the surgical description of his ill-fitting green jacket. Like a steamroller, novelistic prose levels this horizontal, nonhierarchical playing field; and even when it digs deep, the novel moves over surfaces, brushing against them, dusting them, and exposing them, thus showing that true depth lies on the surface, because ultimately, in its irremediably Godless world, there is nothing but surface. Unlike the only other genre that aspires with comparable ambition to be exhaustive—the epic poem—the novel is strictly secular and utterly suspicious of the ineffable and the unspeakable. Methodically descriptive, obsessive in its depictions of the nuances and meanderings of the inner life, and indulgently digressive, the novel is an exhibitionist genre that aspires to use up the universe it creates. It is a genre in which everything washes up to shore and lingers bare-naked, like an unburied body.

Ulysses could be the clearest example (undoubtedly, the most famous and influential) of this exhibitionist strategy and of the epidermal nature of the novel. Joyce himself, along with his friend Stuart Gilbert, even proposed an approximation to the work that seems akin in spirit to the pretenses of this tentative haptic criticism. By identifying each chapter with a part of the body, weren't they encouraging an organic, biological reading and, really, sketching the proprioception of the novel? But a haptic critique of *Ulysses* exceeds by far the aspirations of this essay. A similar approach to certain chapters of *Madame Bovary* (the agricultural fair in Yonville, the carriage ride through the streets of Rouen, the night at the opera when Emma scratches the velvet banister during one of the arias of *Lucia di Lammermoor*) would be a more manageable endeavor. But I would like for now to just give a single example taken from one of the other greatest achievements of the genre.

Moby Dick is the story of a man who sails out to sea in the hope of undergoing a transformation. In the opening passage, Ishmael finds himself wandering the streets aimlessly. He has lost all interest in the world and feels death breathing down his neck. Presently, he stops in front of the window of a funeral parlor and lingers staring at a coffin with morbid melancholy (in the end, he will survive death in the high seas by holding on to a coffin). Ishmael feels that he is fading away, slowly turning into a ghost. He walks to the coast with a heavy heart and a numb spirit. It's a Saturday afternoon and throngs of people from all over town converge by the water. They stand on the boardwalk and look out, mesmerized; Ishmael does too. What is it about water that causes such magnetic fascination in us? Perhaps the answer lies in the ancient myth of Narcissus, Ishmael thinks. Desperate to touch the image on the surface of the water, Narcissus falls in and drowns: "that same image, we ourselves see in all rivers and oceans. It is the image of the ungraspable phantom of life; and this is the key to it all."[22] The decision to set out to sea is a reckless (suicidal even) attempt to apprehend something that will confirm the solid tangibility of the world—something that will allow Ishmael to rebuild himself by recuperating the ability to feel his own self and connect with others. Ishmael's journey is one from the intangible to the tangible. Accordingly, some of the most memorable passages of *Moby Dick* involve the sense of touch. The gelid night at the Spouter Inn, in New Bedford, when Ishmael and Queequeg share the bed and "spoon" in order to fight the cold, or the afternoon when they weave a mat together on the deck of the ship. So too, the digressions on the whiteness of the sperm whale's skin or Ahab's mutilation and his talismanic ivory peg leg are examples of this tactile sensibility. The novel has traditionally attracted allegorical readings, but its pages don't evince the primacy of transcendence, and neither do they privilege the spiritual over the sensible world; they do quite the opposite. What allows for the exegetic somersault of the allegory is precisely the meticulous detail in the description of the craft of the whaler, the candid depiction of life on the *Pequod*, and the compassion and humanity with which Melville portrays his characters. The grand theologico-political vision of the vindictive, monomaniacal leader that guides his followers into perdition, or the narrative of the quest for spiritual salvation in the immensity of existence, only acquires exegetic robustness on the basis of the fastidious and minute depiction of each and every character, of every nook and cranny of the ship, and of the peculiar and idiosyncratic customs of life in the high sea, that Melville presents with an exasperating level of precision. *Moby Dick* demands from us a carnal engagement with the whaling universe into which it casts us. We travel the prose brushing against the words until, suddenly and unexpectedly, as if it were by magic, the rub produces a spark and it all comes to life. Queequeg's tattoos, the baroque deck of the *Pequod*, the multitude of ropes tied in a myriad types of nooses, the foreboding immensity of the whale, the demiurgic figure of Ahab and that of his somber Persian minion, Fedallah. The end comes in the form of the realization of a series of prophecies pronounced by Fedallah. According to one of them, Ahab's death will be preceded by the vision of a hearse.

The hearse is none other than Moby Dick towing the corpse of the very Fedallah, who was caught in the flying harpoon and dragged out to sea. Through this dreadful and sublime image, Melville stages the power of literality and celebrates with flying colors the aesthetic miracle that turns words into tangible realities. But no passage in the novel illustrates this more perspicuously than the beginning of chapter 94.

"A squeeze of the hand" describes the process of removing spermaceti from the depths of the whale's cranium and then treating it to prevent it from congealing. The precious liquid—one of the most profitable products in the whaling industry—was used in the cosmetic industry, it served to tan leather, and it was the raw material for candles, lubricants, and house fuel. After being removed from the whale's head, spermaceti was stored in large tubs and, due to its tendency to swiftly cool down and crystallize, it was necessary to squeeze the slippery and thready lumps back into liquid; this was done by hand. Here is how Ishmael describes the strange pleasure that the task gave him:

> It had cooled and crystallized to such a degree, that when, with several others, I sat down before a large Constantine's bath of it, I found it strangely concreted into lumps, here and there rolling about in the liquid part. It was our business to squeeze these lumps back into fluid. A sweet and unctuous duty! No wonder that in old times sperm was such a favorite cosmetic. Such a clearer! such a sweetener! such a softener; such a delicious mollifier! After having my hands in it for only a few minutes, my fingers felt like eels, and began, as it were, to serpentine and spiralize.[23]

By alternating occlusive and fricative sounds, this passage at once describes and recreates the process of disintegrating the lumps of spermaceti. Like the milky blobs, sounds and images too oscillate between the solid and the fluid. The first sentence, with its predominance of occlusive consonants, illustrates the solidity of the substance that has retreated into itself, that has *cooled, crystallized, concreted.* Immediately after, the process of softening begins with the verb *squeeze,* which combines both types of sounds, and then continues with words abundant in sibilant sounds that recreate the very noise made by the squeezing, exploding, and liquefying of the sperm: *sweet, soft, hands, fingers, eels, serpentine, spiralize.* What makes the substance so unfamiliar to the touch is its liminal physiology, its intermediary state between solid and liquid. We are tempted to imagine it as a bubble bath, but Melville will not indulge us, refraining from any analogies. It is an absolutely new kind of texture that produces a unique sensation in Ishmael, who, in an outburst of joy, praises the novelty and the discovery of something that, until not too long before, would have been deemed impossible: a pleasant duty. Unlike every other task on the ship, squashing the clots of spermaceti is a "sweet and unctuous duty." The simile of the magical metamorphosis by which fingers turn into eels that "serpentine and spiralize" anticipates the transformative effect that the task will have for the men. The passage continues:

> As I sat there at my ease, cross-legged on the deck; after the bitter exertion at the windlass; under a blue tranquil sky; the ship under indolent sail, and gliding so serenely along; as I bathed my hands among those soft, gentle globules of infiltrated tissues, wove almost within the hour; as they richly broke to my fingers, and discharged all their opulence, like fully ripe grapes their wine; as I snuffed up that uncontaminated aroma,—literally and truly, like the smell of spring violets; I declare to you, that for the time I lived as in a musky meadow; I forgot all about our horrible oath; in that inexpressible sperm, I washed my hands and my heart of it; I almost began to credit the old Paracelsan superstition that sperm is of rare virtue in allaying the heat of anger; while

bathing in that bath, I felt divinely free from all ill-will, or petulance, or malice, of any sort whatsoever.[24]

It's clear by now that this is not a mere digression; we have come to a crux in the narrative. Amidst the whirlwind of the whale hunt and from within the confinement of the *Pequod*, Ishmael makes a crucial discovery as he sits cross-legged and exhausted on the deck of the ship, that sails smoothly between a sky and an ocean that are equally blue and equally tranquil. The ocean and the sky extend into the horizon as one immense backdrop without limits or edges, like an all-encompassing bubble that contains the ship and, in it, the tub overflowing with spermaceti in which Ishmael sinks his hands to undo the smooth and silky blobs of whale soap. Protected like a pupa inside a cocoon, impervious to any distractions or preoccupations, impermeable to the outside world, removed from all context, Ishmael fully gives in to the task of kneading and bursting the "soft, gentle, globules of infiltrated tissues" that explode, juicy and opulent like grapes. The unpolluted substance exudes the smell of spring violets and Ishmael is transported to a "musky meadow," that acts both as *locus amoenus* and baptismal font, a place where the soul finds its rest and where evil has no business. As his hands rub the clots of sperm, another set of invisible hands massages Ishmael, ridding him of the tension produced by the oath of hatred and revenge that leads the ship toward its unfortunate and imminent fate. The bath in whale sperm is a feast for the senses of smell and touch, but its bliss transcends the sphere of mere epidermal pleasure; it drains inward, into the body, through the pores, and washes off the stains of perfidy and malice from Ishmael's obfuscated interiority. And so we come to the climax:

> Squeeze! squeeze! squeeze! all the morning long; I squeezed that sperm till I myself almost melted into it; I squeezed that sperm till a strange sort of insanity came over me; and I

> found myself unwittingly squeezing my co-laborers' hands in it, mistaking their hands for the gentle globules. Such an abounding, affectionate, friendly, loving feeling did this avocation beget; that at last I was continually squeezing their hands, and looking up into their eyes sentimentally; as much as to say,—Oh! my dear fellow beings, why should we longer cherish any social acerbities, or know the slightest ill-humor or envy! Come; let us squeeze hands all round; nay, let us all squeeze ourselves into each other; let us squeeze ourselves universally into the very milk and sperm of kindness.[25]

Ishmael carries out the task intoxicated with physical and mental pleasure. As each blob of spermaceti dissolves, time and space dissolve as well. Like an embryo floating in amniotic fluid (blind, nothing but skin, pure movement), the narrator goes into trance and begins to dissolve himself in the substance. Squeeze! Press! Squash! Squeeze until you melt! Now it is not just his fingers that turn into eels, it is his whole body that serpentines and throbs dipped in the magical substance, a cluster of raw limbs revealing a shredded, hypersensitive subjectivity as receptive as a gaping wound. And when he is about to come completely undone and be lost in the pearly substance, he finds his comrades' hands—hands that, like his own, are hard to distinguish from the lumps of spermaceti. At this point, the eels recuperate their original form; they go back to being hands and fingers, and the moment of redemption occurs. Deep in the foam of the whale's essence, the task becomes a ritual of communion, a liturgy of friendship, of affection and love: a mystical union. This unspoken pact of fraternity, founded upon physical contact and affect, replaces the oath of hatred, violence, and revenge demanded by Ahab as token of allegiance that functions as the de facto principle of community onboard the *Pequod*. Until that moment, the visual had been elided, but all of a sudden it comes back as the men look up and, both moved and surprised, confirm that they had not been squeezing their hands by mistake. And, for a second, the cloud of hardship and envy, of loathing, death, and destruction that loomed large in the voyage dissipates as the men find in one another a companion; they greet, they hold hands, and they look around as if they had never before laid eyes on each other. The rediscovery of affect in shared duty and in tactile pleasure ignites in Ishmael a deep longing to become one with his fellow men and to dissolve into what is most basic and primordial in man, into what Melville calls "the very milk and sperm of kindness." But this will never happen. As soon as they are done with the task, they will once again grow apart and fall under the odious spell of the captain. It is hard not to read into this a deep political pessimism. Love, fraternity, and equality among human beings are possible only in close proximity, when there is contact and immediacy. This, of course, can happen but it is as sporadic as it is fleeting. The passage concludes with a moral that would sound trite and anodyne (as morals usually do), were it not for the strange image with which it ends:

> Would that I could keep squeezing that sperm for ever! For now, since by many prolonged, repeated experiences, I have perceived that in all cases man must eventually lower, or at least shift, his conceit of attainable felicity; not placing it anywhere in the intellect or the fancy; but in the wife, the heart, the bed, the table, the saddle, the fire-side; the country; now that I have perceived all this, I am ready to squeeze case eternally. In thoughts of the visions of the night, I saw long rows of angels in paradise, each with his hands in a jar of spermaceti.[26]

If you are governed by ideas and lost in the pursuit of chimeras, the forest won't let you see the trees, Ishmael concludes. The only attainable form of happiness is that found in the small sensorial and affective pleasures: a loving touch, a soft bed to rest our bones in after a hard day, a bite of something delicious, a warm spot by the fireside. We might add the pleasure of reading, which oftentimes, when the vicissitudes of everyday life make us numb, invigorates our ability to feel and to be affected by allowing us to get lost in the narrative and by opening us to the reality of empathy. Writing too can produce such raptures. In a letter to his friend Richard Henry Dana Jr.—the earliest-known reference to *Moby Dick*—Melville compares writing with extracting oil from whale blubber. On May 1, 1850, Melville writes from New York: "As for the 'whaling voyage' I am half way in the work. . . . It will be a strange kind of book, tho', I fear; blubber is blubber you know, tho' you may get oil out of it, the poetry runs as hard as sap from a frozen maple tree; and to cook the thing up, one must needs throw in a little fancy, which from the nature of the thing, must be ungainly as the gambols of the whales themselves."[27] Hands undoing blobs of sperm, squeezing out the valuable oil, hands transformed into slippery eels, hands that touch, hold, and caress other hands, recognizing them as equal, constitutive parts of a whole that transcends them: all these are examples of that ungainly fantasy that runs through the novel. The novelist Herman Melville extracts the nectar of poetry from the dense prose and, in doing so, shakes our hand. Appropriately, his muses are long rows of angels each with his hands in a jar of spermaceti.

SIX FINGERS

In the beginning there was Homer. And Homer spoke about God. And Homer spoke *with* God. But in the beginning God was gods. Hesiod says that mortals and immortals are all the children of Gaia, of mother earth—Pindar later confirms it. We belong to the same race and are moved by the same passions, but the gods will not die, whereas we will. We get sick, we grow old, we wear out; they do not. They dwell in the sky, we live on earth. In order to give a sense of the distance that separates us, Hesiod says that it would take nine days and nine nights for a free-falling anvil to get from their abodes to ours. And yet, from distant Olympus, the immutable gods extend their arms and reach out for us. They demand sacrifices, libations, and prayers from kneeling suppliants. They delight in the ephemeral tragedy of human life sung by inspired poets (they might not know it, but their blissful lives too beat to the rhythm of the dactylic hexameter). On occasion, they even fall in love with men and women and spawn demigods and heroes, perhaps driven by that bizarre tendency that makes superior natures become mesmerized by inferior ones—a tendency that, according to Plotinus, explains sin. Or perhaps because we are all the children of Gaia, we are all part of the cast of this never-ending epic. Whatever the reason may be, the gods, from their unfathomable distance, approach us. This is one of the wonders that Homer strives to account for in the *Iliad*.

In the beginning there was Homer, but Homer was not too

interested in beginnings. A couple of generations later, the muse would inspire Hesiod to start from the beginning. What followed was that archeological fixation that would dominate the poetry of the following centuries and that would give birth to philosophy, physics, and history. Unlike Hesiod, Homer is not too concerned with theogonies, or with the origins and causes of human suffering. Toward the end of the *Iliad*, Achilles expresses this disinterest through the parable of the two jars—one filled with miseries, the other with blessings—from which Zeus randomly takes to hand out to mortals. At the end of the day it makes no difference whether the cause of the war in Troy was the kidnapping of Helen and whether the gods favor a specific warrior out of spite, or in gratitude for their piety. What matters for Homer is the conflict itself, the *agon*: Chryses versus Agamemnon, Agamemnon versus Achilles, Achilles versus Hector, Hector versus Athena, Athena versus Apollo, Apollo versus Zeus, Zeus versus Hera. What matters for Homer are the passions that any given conflict stirs: Achilles' rancor, Agamemnon's arrogance, Diomedes' fury, Nestor's nostalgia, Dolon's terror, Hector's patriotism, Priam's sadness. Emotions are the flesh and the blood of the poem. Devices such as speeches and epithets, similes and catalogs, digressions and repetitions are its muscles, joints, and tendons, whereas the plot is its skeleton. So is the fabric of the colossal body of the *Iliad*.

The poet casts us without notice into the boiling cauldron of the action, in medias res. The immediacy of the in medias res produces confusion and disorientation. We have landed on an unknown continent; we are caught up in the middle of a nine-year-long conflict and are denied the proper perspective to find our bearings. Orienting ourselves spatially in the landscape ranges from difficult to close to impossible. All we have is whatever lies at arm's length, vibrating to the beat of the six dactyls: a simile, an epithet, a speech to which we can hold on, like a buoy in a shipwreck, in order to rest, breathe, and move forward. We navigate the Homeric space following the rhythm of the dactyl and we travel through time led by Dawn, whose "rosy fingers" turn the pages of the days.

We know little to nothing about Homer's predecessors, except for the fact that the Homeric poems belong to a long tradition of oral poetry that gets lost in the haze of the dark ages. For all intended purposes, the dactylic hexameter is born with the *Iliad*. The meter consists of six feet, dactyls (one long and two short syllables), which can occasionally be substituted by spondees (two long syllables). However, among the great classical epic poets, Homer is the one who least indulges in such substitution; the dactyl is his preferred foot. Dactyl means "finger," the three syllables stand for the three phalanges of four of our fingers (the binary thumb representing the odd spondee). This link between meter and the shape of the hand seems to point back to the irretrievable, quasi-mythical dawn of poetry itself. Meter determines the rhythm, the accents, the caesuras, the intonation, and the very performance of the poem; in this sense, it is a crucial component of the creative process. We know virtually nothing about the composition of the Homeric poems, but it is clear that the impositions of the meter played a fundamental role in it. The law that rules the poem is neither that of Zeus nor that of Homer: it is the law of the dactyl that balances the internal structure and determines the way in which the lines ought to be sung. The metrical fingers, after shaping the formal surface of the poem and determining most word choices, stretch from the vocal chords of the *aoidos* until its tips reach the ears and, through them, the heart of the listener; these fingers might be invisible, but they are by no means intangible. The dactyls thus equip the poem with exteroception and intraception. As they shape its formal structure, they give it its own balance and, by making it into an organic whole, they endow it with its own proprioception. Finally, by regulating its cadence and speed they bestow movement on it: they grant it the gift of life. If passions and emotions are the flesh and the bloodstream of the epic poem and the plot its bones, tendons, and ligaments, the meter is its haptic apparatus.

Homer does not spare dactyls; he avoids substituting them; he wants them all there performing the great symphony of the epic, stretching the poet's tongue, pulsating, tapping, scratching, and caressing the affective nucleus of the audience. The intimacy produced by dactyls, along with the constant sense of immediacy, generate a form of narrative that unfolds mostly in the short and middle distance. Not unlike the Achaeans and the Trojans, we too spend most of the *Iliad* trapped, stuck in a war that has gone on for too long, locked inside a city under siege, caught in the middle of battling forces. The moments in which the poet offers panoramic views ("establishing shots" one would say in the language of film) are rare, and they tend to occur when, through the intervention of a god, Homer travels the daunting distance from Olympus to the earth. For the most part, the dactyls keep us close to the ground, in the midst of the conflict. The *Iliad* not only begins in medias res, it also unfolds in medias res: we are always in the middle of something. We do not see much. The journey is haptic rather than visual.

This brings me back to Riegl and to his distinction between optic and haptic cultures. Sweeping though the categorization may be, and in spite of the Hegelian (and decidedly obsolete) set of assumptions that any such historiography betrays, Riegl's notion of a haptic and an optic *Kunstwollen* still proves stimulating. But because either option would lead to a dead-end road into the past, the only alternative is to combine them and to transform the dichotomy into a contradiction. Homer's poetry is haptic and it is not haptic; it is optic and it is not optic. While for the most part the poet drags us through the dusty battlefield, among men who debate or fight to the death, at times he also elevates us. He positions us on the walls of Troy, from where we review the troops along with Helen and Priam. Or he takes us from the earth to Olympus to witness the gods' agitated negotiations, and then back to earth faster than a free-falling anvil. Like the tides of the wine-dark sea, Homer's dactyls ebb and flow, zoom in and zoom out, expand and then recede. Like the Olympian gods, Homer moves from the imperturbable beyond to the heat of full physical contact (bodies, blood, tears, sweat, dust, bronze, and leather) in the space of one hexameter.

We can begin with Apollo, god of poetry and bringer of the plague. Apollo's shafts start the epidemic that ignites the main conflict of the poem, the power struggle between Achilles and Agamemnon. The Achaeans have kept a fruitless siege on Troy for nine years, and in a matter of nine days Apollo speeds the outcome of the war by disseminating pestilence and death among the invading armies, their cattle, and their dogs. Apollo's ferocious *deus ex machina* comes in the first sixty lines of book 1, and soon remains overshadowed by the pedestrian and gaudy dispute between the two warlords. Apollo descends from Olympus, urgent and inexorable like the night, to attend to the prayers of Chryses, his humiliated priest. But it is actually the poet who summoned him, so that he would start the engine of the epic with his venomous darts. One of the preferred epithets for Apollo is "he who shoots from afar" (*hekebolos*, or *hekatebolos*). Leto's son, honoring his epithet, takes position at a distance and, kneeling by the ships, shoots one shaft after another for nine long days, inoculating the army with a deadly disease. Homer also calls the god *hekaergos*, "he who works from afar." And yet Apollo does nothing but come close: first, as he travels swiftly from Olympus to earth, and then, by means of his poisoned arrows, as he enters the bloodstream of the Achaean armies. Apollo jumps from the faraway peaks of Olympus, from where one can see the world with the utmost clarity, to the battlefield and, from there, by means of his arrows, to the blood and the guts of men, the dark world of the inner body where nothing is visible and everything is tangible. This abrupt pattern of movement by the god of poetry, from far distance to the closest proximity, initiates the action of the epic and reflects Homer's *ars poetica.*

The *Iliad* is a succession of affective explosions in a desert of prolonged delays, standstills, and inaction. The spatial ambiguity and the constant feeling of disorientation are structural elements of a narrative whose main setting is a wasteland adjacent a city under siege. The tedium and the claustrophobia after nine long years of being trapped (the Trojans in Troy, the Achaeans in their camp) hamper sight, and the surrounding space becomes indistinct and blurry. Beyond the fact that Troy has high walls and wide streets, and that the Achaeans have built ramparts (that infuriated Poseidon), beyond the odd mention of a river, a rock, or a promontory, we know little to nothing about the space where the action unfolds. The narrative is, instead, located in the uncharted territory opened by the constant outbursts of emotion. The rancor, wrath, and pain of Achilles, the anguish of Priam, Andromache and Hecuba's despair, Astyanax and Dolon's fear, Diomedes' fury, Nestor's cold serenity, etcetera, guide us through the hazy terrain of the action. The poem creates space and opens ways through it by means of haptic, rather than optic, elements: the hands of suppliants that hold knees and stroke beards, the shedding of tears, outbursts of rage and pain, spears that pierce skulls, bladders, and livers, even Achilles' fingers, idly striking the chords of the lyre while revealing with elegant decorum the intensity of his spite. For the most part, it is the mechanism of request, or deliberation, followed by violent action that drives the plot forward. This is regularly interrupted by Homer's notoriously elaborate similes, whose objective is not just to give the audience a break, albeit brief, from the brutality of the action, but also to offer a visual oasis. Similes transport us to worlds where we do orient ourselves, worlds that we do know, the world of hunting and tending for cattle, the world of sailing and making cheese. Scenes from everyday life, such as children disrupting a wasp nest, a little girl clinging to her mother's dress crying, or a dog begging for food by the table, shine like flashbacks of the world that the warriors have left behind nine years before and where we, the audience, are presumably positioned. One simile in particular, both striking and unusual, is especially provocative and revealing about the haptic nature of the poem. It comes during the most extreme moment of tension: the struggle over Patroclus's body.

Book 17 deals with the fight between Achaeans and Trojans to recuperate the body of Patroclus. In the previous books, the escalation of violence has turned men into wild animals. Even Hector, who upon leaving Troy in book 6 was a compassionate family man determined to put an end to the war without any further loss of life, has become, in the heat of battle, like a bloodthirsty lion that foams from the mouth and slaughters Achaeans without any mercy. Diomedes, Ajax, and the very Agamemnon are wounded after killing large numbers of enemies. In book 16, as the Trojans threaten to burn the Achaean ships, Patroclus puts on Achilles' armor and runs to assist his comrades. In one of the most dramatic scenes of the poem, Patroclus dies at the hands of Hector. The struggle for his corpse brings out the most primal instincts in the soldiers. Menelaus is compared to a bloodthirsty lion when he murders Euphorbus, and again to a lion surrounded by dogs when he protects Patroclus's body from a group of Trojans. Ajax too is compared to a lion defending a litter of cubs, as he protects Patroclus from Hector, who having already stolen his armor is trying to drag the body away in order to behead it and throw the head to the dogs. Violence erupts again when Hippothous, as he manages to get a hold on Patrolcus's foot and starts trying to drag it toward himself, is killed by Ajax's spear. Hector strikes back with his own spear, which brushes against Ajax, continues its trajectory, and kills Schedius. Ajax then attacks Phorcys, who drops dead on top of Hippothous's corpse, pierced through the stomach, his intestines falling out. The heat of the battle produces a state of haze and blindness. Vision is almost nullified, and it is impossible to say whether it is day or night. It is an utterly haptic moment. In Homer's words: "So on they fought like a swirl of living fire—you could not say if the sun and moon still stood secure so dense the battle-haze that engulfed the brave who stood their ground around Patroclus' body."[1] The struggle goes on for hours and the ground darkens as it absorbs the blood of piles of dead men. The heroes are exhausted, the gods have had enough, and the poet is drained. Tension has reached its breaking point and Homer illustrates it with a stunning simile:

> So all day long for the men of war the fighting raged,
> Grim and grueling, relentless, drenching labor, nonstop,
> And the knees, shins and feet that upheld each fighter,
> Their hands, their eyes, ran with the sweat of struggle
> Over the great runner Achilles' steadfast aide-in-arms,
> An enormous tug-of-war. As when some master tanner
> Gives his crews the hide of a huge bull for stretching,
> The beast's skin soaked in grease and the men grab hold,
> Bracing round in a broad circle, tugging, stretching hard
> Till the skin's oils go dripping out as the grease sinks in,
> So many workers stretch the whole hide tough and taut—
> So back and forth in a cramped space they tugged,
> Both sides grabbing the corpse and hope rising,
> Trojans hoping to drag Patroclus back to Troy,
> Achaeans to drag him back to the hollow ships
> And round him always the brutal struggle raging.
> Not even Ares, lasher of armies, not even Athena
> Watching the battle here could scorn its fury,
> Not even in their most savage lust for combat, no—
> So tense the work of war for the men and chariot-teams
> That Zeus stretched taut across Patroclus this one day . . .[2]

The space has dramatically shrunken and the soldiers are caught within a bubble of toil. Their feet, legs, and knees are soaking with sweat; their hands are caked with dirt and blood, and they are blinded, wrapped in a cloud of dust. But there is also a metaphoric blindness at play; the increasing fatigue, alongside an obsession with winning the tug-of-war, also blinds them. Numb and filthy, the men pull from the corpse like hyenas fighting for a piece of carrion. All signs of humanity are lost, or suspended, and the gods have no intention of intervening; we are in no man's land, under the sole rule of brute force. The story of Achilles' rage has suddenly been stripped of every tragic underpinning, of all nuance and of all mythology, and is encapsulated inside a bubble of dust and perspiration where there is nothing but tactility: the drops of sweat that run down the soldiers' legs, the dirt that covers their hands and their eyes, the desperate, furious grip onto the corpse's ankle, onto his wrist, onto his mane, the exhaustion that macerates their muscles, the extreme tension of the tug-of-war. This grotesque image, perhaps better than any other, captures the *Iliad* for what it is: a song about anger, fear, and exhaustion. The simile that Homer uses to illustrate this is conveniently tactile; it transports us to the world of tanners. A group of men vigorously stretch the hide of a bull lathered in fat under the Ionian sun. Through the tension of the stretch, the hide releases its moisture and absorbs the fat becoming incorruptible, resilient, leather.

The most immediate meaning of this simile points to the sacrilegious treatment of Patroclus's body, who, instead of receiving the proper funeral rites worthy of a hero, is manhandled as if he were an animal product. But under this layer of exegesis there are others. The tanners pulling from the stretched hide do not simply refer to the soldiers fighting a tug-of-war over the corpse; their toil also reflects the peak of tension reached in the narrative. With all that was left of humanity in the soldiers gone, and with the gods out of the picture, Homer reduces the space to the minimum, interrupts time, and stops the action; all that is left is pure tension—a tension that feeds back into itself and that sustains itself in a state of stagnation that is, paradoxically, both grueling and frantic. It is precisely through this suspension of action, of time, and of space that a third sense of the simile is revealed which has nothing to do with the specifics of the struggle over the dead body, or with the overarching story, but that opens a way to think about the haptic nature of the poem.

The *Iliad* tells the story of Achilles' anger, an emotion that, at first, surfaces as deep resentment against Agamemnon and then, after the death of Patroclus, as full-blown rage against Hector. The initial rancor is deadly. Achilles goes on the sidelines and allows the Trojans to decimate the Achaeans. His subsequent rage is even deadlier. Armed by his mother, the son of Peleus carries out an anthological massacre of Trojans by the walls of the city and in the banks of the river. The struggle over Patroclus's body, during which the Trojans steal his armor and his weapons, is the immediate cause of the transformation of Achilles' resentment into rage. Book 17 focuses on this struggle, but it is traversed through and through by the ghost of Achilles, who still lingers in his camp ignorant of the fate of his beloved companion. The Achaeans debate in fear about who should break the news to the hero. The terrified Trojans anticipate his return to the battlefield. The stress of the war and the tension caused by the ever more palpable anticipation of Achilles' imminent outburst, find a magnificent symbolic instantiation in the tug-of-war and in the simile of the tanners. The direction of the poem is about to change dramatically; we are headed for the final stretch. Not unlike the tanner, the poet has his men push the action to its limits so that he can begin to sing the conclusion. The hide of the bull loses its perishable qualities as it releases moisture and dries in the sun; the feats of the warriors, which could easily be lost in oblivion, like forgotten or neglected corpses on a battleground, are immortalized by the epic poem.

In the fifth book of the *Histories*, as he discusses the Ionian rebellion against the Persians, Herodotus says that "the Ionian term for papyrus rolls—namely 'skins'—also goes back a long way, to when they used goatskins and sheepskins to write on, because they did not have any papyrus. In fact, even today many non-Greeks use such skins for writing."[3] Patroclus's body—scapegoat of the gods, of the muse, of Achilles, of Homer, of the *Iliad*—is the text that immortalizes the great feats of yore by the walls of Troy. In book 19, Achilles makes peace with Agamemnon and says to him: "For years to come, I think, [the Argives] will remember the feud that flared between us both."[4] The hide has been tanned and its dry, impermeable, enduring texture is ready to become text. This obsession with storytelling as the only tangible form of immortality (of transcendence) is typical of the Homeric epic. Helen, mythical cause of the war, had already repined the gods and their sadistic eagerness to sow misfortunes among humans so that poets later sing about them: "Zeus planted a killing doom within us both, so even for generations still unborn we will live in song."[5]

The first impression on the blank parchment is a scream of pain. Antilochus, chosen by the Achaeans to be the bearer of the news, arrives at Achilles' tent and informs him of the loss. Achilles' scream in the beginning of book 18 perforates the capsule of tension that had formed over the battlefield. Time stands still once again, but now the space has opened, dilated by the piercing scream of mourning. And then Achilles screams again. His howls travel through the battleground like invisible Furies. They freeze the hearts of Achaeans and Trojans alike, and they reach the nymph Thetis, who rushes to her son and promises him a brand-new armor. Thetis proceeds then to pay a visit to Hephaestus, the god of smiths and of craftsmen, and this is how we get to one of the most memorable passages in the history of Western literature.

The shield that the god makes for the hero is described in a famous ekphrasis that takes up most of the final part of book 18. The ekphrasis, a verbal illustration of a visual artifact, is a literary device that establishes a dialogue between two forms of art and that, therefore, invites a reflection on the nature of art itself. But there is more. In Homer, the ekphrasis, not unlike the simile, acts as a visual oasis where we rest from the all-too-tangible quagmire of the war, and thanks to which we orient ourselves in the confusing geography of the conflict by means of the appeal to worlds that we do know. The ekphrasis rescues us from the plot like a *deus ex machina* and, as it elevates us, it provides new vantage points and novel perspectives. Like the tanning simile, the description of the shield of Achilles, as it creates distance and incites the eye, also invites a reflection upon the haptic nature of poetic creation.

The description of the shield made by Hephaestus is one of the most singular moments in literary history for a number of reasons. First, in all of its detached and contemplative serenity, the long excursus causes a violent swerve in the narrative. After eighteen long books spent mostly close to the dusty ground, in the midst of the action, and in the promiscuous human clutter of the Achaean camps, the poet elevates us. Very few moments until then (namely, some similes, a couple of episodes in Troy and among the gods in Olympus) had removed us, always briefly, from the action. But even then the war remained an inescapable reality imposing itself from all sides, either discussed by the gods in heaven or witnessed in awe from the high walls of the city. To peruse Achilles' new shield, however, feels like flying away from Troy to a different world, a perfectly quotidian and familiar world of weddings and street festivals, of commerce and tribunals, of shepherds and winemakers, of kings and peasants, of oxen, lions, and lambs. There are two cities in this faraway world, one that lives in peace and prosperity, and next to it the god carves another one that is at war and under siege, another Troy. But not even this vision drags us back to Priam's Troy. As we contrast this troubled city with the other one, where men and women dance and sing, drink and love, trade, litigate, and negotiate with civility, far from being thrown back into the action, our detachment increases and we appreciate the totality of the world that has been there, on the horizon, all along. In doing so, we are momentarily freed from the haptic entanglement of the *Iliad*; we open our eyes and see, we catch our breath.

The ekphrasis is even more singular, though, because in order to elevate us the poet introduces us in the heart of Hephaestus's forge, a metallic world of ovens, tools and cauldrons, anvils, hammers, and bellows. The aerial vision that reveals a whole cosmos carved on the shield—a sight that, according to the lame god, will astonish humans for generations—is born in the confined and glowing space of the smithy. And it is the very carving of the shield, the craftsmanship itself that gives the ekphrasis its most singular characteristic. Leaning over the anvil, hammer in one hand and tongs in the other, the god sets out to produce a shield and ends up creating the world. The verb Homer uses is *poiein*, "to make," from where the words poet and poetry stem. In the case of the blacksmith, to make means to carve, and to carve is to turn two dimensions into three. Hephaestus's creation is a metonym for artistic creation; and it is eminently haptic.

> At first Hephaestus makes a great and massive shield . . . and across its vast expanse with all his craft and cunning the god creates a world of gorgeous immortal work. There he made the earth, and there the sky and the sea and the inexhaustible blazing sun and the moon rounding full, and there the constellations, all that crown the heavens. . . . And he forged on the shield two noble cities filled with mortal men. With weddings and wedding feasts in one and under glowing torches they brought forth the brides . . . and the young men came dancing, whirling round in rings . . . women rushed to the doors and each stood moved with wonder. . . . But circling the other city camped a divided army gleaming in battle gear. . . . And he forged a fallow field, broad rich plow land tilled for the third time. . . . And he forged a king's estate

> where harvesters labored, reaping the ripe grain, swinging the whetted scythes. . . . And he forged a thriving vineyard loaded with clusters, bunches of lustrous grapes in gold, ripening deep purple and climbing vines shot up on silver vine-poles. . . . And the crippled Smith brought all his art to bear on a dancing circle, broad as the circle Daedalus once laid out on Cnossos' spacious fields for Ariadne the girl with lustrous hair. Here young boys and girls . . . danced and danced linking their arms, gripping each other's wrists. . . . A breathless crowd stood round them struck with joy and through them a pair of tumblers dashed and sprang, whirling in leaping handsprings, leading on the dance. And he forged the Ocean River's mighty power girdling round the outmost rim of the welded indestructible shield.[6]

Like a magician, the god of weathered hands and lame feet performs the miracle of creation for Thetis's eyes: the metal sheets begin to rise, their plain horizontality curves and acquires relief, their two dimensions become three. The god's hands (tools of tools, as Aristotle would say) have produced something more real than the combination of bronze and gold. And the sad countenance of the nymph—that pagan *Pietà*, mother of Achilles "who would not live long," as W. H. Auden says in his dystopian rendition of the scene—lightens with awe as the metal rises, folds, and becomes engraved. All of a sudden, we have the earth, the sky, and the sea, the midday sun, a full moon, and all the constellations that shine in the distance, like minuscule stains splattered on the dark canvas of the universe. Two cities then emerge. In the first one, there is a wedding celebration with torchlight processions and people singing epithalamia (there is poetry, there is music). Groups of young people dance to the melodies of flutes and lyres (there is movement, there is dance). The metal is no longer metal; it is a poem. Thetis forgets for a moment the tragic fate that awaits her son and is lost in the merry action; poetry can do this. Poetry can also stage horror and disaster. The second city that emerges from the metal sheet, like a volcanic island in the morning of the universe, is reminiscent of Troy: there is a siege, people are trapped, the invading army lingers, there are assemblies and ambushes, betrayals, blood, and corpses being dragged through the dust.

But not all is urban in the cosmos of the shield. The god also "adds" (or "puts": the Greek verb is *tithemi*) fertile lands, peasants, and plows. This notion of "adding," or "putting," on the metal, instead of drawing from it (for Michelangelo, sculpting consisted in bringing the pre-existing form out of the block of marble) shows that the god's craftsmanship is conceived as an act of magic, which makes something appear out of nothing. To the Greeks, there is no creation of the world; the world has always been there and it will always be there. But there is artistic creation. And artistic creation is, first and foremost, a manual skill. The god uses the hammer, forces the metal imposing his craft on it, and manufactures the shield. It is only then that the other senses are summoned. First, sight and sound, and after, taste and smell: after a long day of work with the plow, peasants drink wine (there is flavor, there is fragrance). The aromas intensify and mouths begin to water when, not far from there, the god "puts" the orchards and gardens of a king, promptly tended by servants who can already smell the feast that awaits at the end of a hard day of work: an ox offered to the gods, its meat seasoned with barley and salt. And in this very estate, the god "puts" exuberant vineyards filled with lush black bunches of grapes ready to be picked, carried in wicker baskets filled to the brim by young boys and girls who sing and dance, jump and cheer. The creation involves movement and the movement accelerates and intensifies because the lame god also "puts" there a dance floor similar to the one designed for Ariadne by Daedalus in Cnossos, the poet adds, where girls in golden tresses wearing linen dresses and boys in immaculate tunics hold each other from their wrists and dance around frantically. They dance and swirl like the wheel of a potter, the poet says. Around them, also in a circle, a crowd of onlookers is delighted by the spectacle. Both circles are, in turn, surrounded and contained by the last element of the blacksmith's creation, the Ocean River that girdles round the outmost rim of the shield. This wonder of concentric circles is reflected in the eyes of Thetis, who has witnessed the process of creation and is now faced with the finished product. The image of the concentric circles also bespeaks the nature of poetic creation for Homer: stories within stories, flashbacks and flash forwards, distant worlds alluded to through similes, the world of mortals and the world of the immortals, the whole spectrum of the tragicomedy of human life in twenty-four books. The comparison between the circle of dancers and the potter's wheel concludes the ekphrasis with artisanal mastery. Artistic creation is born in the hands, and its biggest accomplishment is movement. Toward the end, Homer mentions Daedalus, the mythical Cretan sculptor who, according to Plato, carved statues that were so alive and eager to move that they had to be put in shackles to prevent them from escaping. In the remote myth of Daedalus, the figure of the artisan and that of the magician are one and the same.

There is another Greek myth that concerns craftsmanship and magic. According to an old legend, when Rhea became pregnant with Zeus, and in order to prevent Cronus from eating the child, she fled to Crete. Once there, she found a mountain (some say Aigaion, others claim it was Dikte, or Ida) and hid inside a cave. When the time came to give birth, the goddess squatted and, resting on her arms, sunk all ten fingers on the dusty ground. This is how Zeus was born and this is how another race of deities sprung to life, because, from the imprint of her fingers, ten dactyls were born, five boys and five girls. The males are blacksmiths, the females are magicians, and their names are among the best-kept secrets of ancient Greek religion. Robert Graves and Karl Kerényi agree that the ancient myth of the dactyls refers to the origins of metallurgy and to the magical and symbolic dimension of the figure of the blacksmith.[7] It reveals the sacred respect in which the handling of metals, and skills like that of the goldsmith and blacksmith, were held during the Bronze Age. Ultimately, I may add, the myth of the dactyls is evidence of the veneration humans have always had for their own hands.

A few centuries after Homer, Anaxagoras would claim that the distinctive note of human beings, what separates us from all other animals and guarantees our dominion over them, is the sophistication and dexterity of our hands. With our hands we make tools and forge weapons, we build houses, we weave clothes, we write, play music, and make art.[8] Many years before Homer (about thirty-eight thousand, give or take), in the cave of Leang Timpuseng, on the island of Sulawesi in Indonesia, anonymous artists left imprints of their hands stenciled in ochre on the walls. The Leang Timpuseng hands are to this day the oldest-known cave painting in the world, and, as such, the earliest-known work of art. As in the story of Rhea, the cave appears as a sacred space—paradigmatic microcosm, metaphor for both intrauterine existence and the afterlife—and hands as the symbol of human astonishment and reverence in the face of their own skill.[9] Many millennia after the Upper Paleolithic artists (Were they in fact women, as some have argued?) left their imprint on the cave walls, in Renaissance Rome, Michelangelo focused the moment of creation of the world in two fingers that have either already touched or are about to touch. This primal connection between artistic creation and hand dexterity is also at the basis of the Homeric ekphrasis. The five male dactyls are patrons of metallurgy, the five female, of magic; together, they account for Hephaestus's wondrous creation, the shield of Achilles. As for poetry, it is six dactyls that mark the rhythm of the epic. Like the god, the poet also creates movement and life and does so by handling and molding the dactyl with similar artistry.

In the first chapter of *Mimesis*, Auerbach claims that the Homeric poems and the Old Testament are the two foundational pillars of Western narrative. His argument constitutes one of the most-cited tropes in the literary criticism of the twentieth century. What distinguishes the Homeric style from that of the Old Testament, says Auerbach, is that Homer is mainly concerned with externalizing feelings, with illustrating scenes as comprehensively as possible, and with meticulously portraying his characters. The narrative of the Old Testament, on the other hand, purposely obscures and breaks up the story in order to direct the focus toward specific moral teachings, always creating an atmosphere of ambiguity that stimulates our dread and awe in the face of the divine. Homeric poetry, Auerbach concludes, takes us to a "realm where everything is visible."[10] Interestingly, the title of the chapter, "Odysseus' Scar," refers to the famous scene of anagnorisis in book 19 of the *Odyssey*, when Eurycleia, Odysseus's old nurse, recognizes her long-estranged master as she washes his feet and legs. It is interesting, I say, because this is one of the most explicitly tactile moments in Homeric epic, and in ancient epic for that matter: the nurse realizes that the man in front of her is Odysseus when she touches his scar.

The Eurycleia scene shows the epistemological relevance that the sense of touch had for Homer. The nurse is the only character in the poem that recognizes Odysseus (human character that is, because he is also recognized by his old dog Argos). Through his uncanny ability to conceal his identity, the hero manages to always have control over any given situation. He does it when he reaches the land of the Phaeacians, he does it in order to escape from the man-eating Cyclops, and he does it when he returns to Ithaca to put an end to the perilous conspiracy of the suitors who are plotting to steal his wife and his throne. In every other occasion, Odysseus reveals his identity only when he sees fit, having the upper hand until the end. Neither Eumaeus, the faithful swineherd, nor the very Penelope are able to recognize him when he presents himself in rags as a wandering beggar. The nurse, however, as she is given the order to wash the stranger's feet, has a sneaking suspicion: "Listen to me closely, mark my words. Many a wayworn guest has landed here but never, I swear, has one so struck my eyes—your build, your voice, your feet—you're like Odysseus . . . to the life."[11] All of her senses are telling her it's him, but only when she kneels before him and proceeds to rub his calf (the Greek verb is *epimassein*), she touches the scar and knows. The tactile evidence is indisputable. Odysseus's emblematic wit and cunning do not stand a chance against the familiar touch of his old nurse: "That scar—as the old nurse cradled his leg and her hands passed down, she felt it, knew it, suddenly let his foot fall. . . . Yes, yes! You are Odysseus—oh dear boy—I couldn't know you before . . . not till I touched the body of my king."[12]

Chapter 27 of *Genesis* offers a fascinating contrast with this episode. Jacob pretends to be his older brother Esau, in order to get his dying father's blessing. In order to fool Isaac, Jacob wears his brother's clothes and, following his mother's advice, covers his arms and his hands with the skin of a lamb. When he approaches Isaac, the old patriarch says to him: "Come close, pray, that I may feel you my son, whether you are my son Esau or not."[13] Upon touching him, Isaac becomes convinced that it is his firstborn: "The voice is the voice of Jacob and the hands are Esau's hands."[14] Thus, giving more credence to his touch than to his hearing, Isaac blesses Jacob. While Eurycleia trusts touch and discovers the truth, Isaac goes by touch and is deceived. Unlike the Homeric poem, the Hebrew text seems to be a cautionary tale regarding the perils of blindly trusting a low, animalistic sense over a high one, like hearing. In both cases, however, touch is presented as the *non plus ultra* of sense perception. For Homer, though, touch is also an ontological *non plus ultra*.

In book 11 of the *Odyssey*, the errant hero performs a ritual to summon the dead. Circe has assured him that only after doing this he shall discover the details of his return to Ithaca. The episode is considered to be the first catabasis, or descent into the underworld, of Western literature; only Odysseus doesn't really descend. If anything, the descent takes place, as some have argued, in his own interiority, although Homer (whose sensibility is as alien to the subtleties of allegorical thinking as can be) does not suggest that Odysseus undergoes a spiritual transformation, like Aeneas will in book 6 of the Virgilian epic. Following Circe's instructions, Odysseus digs a hole in the ground and invokes the dead by offering a series of sacrifices. The fumes of charred meat and the stench of black blood lure the ghosts from the underworld. Once on the surface, they drink the blood offered by the hero and only then can they talk. In Odysseus's world, the dead can see, hear and speak, smell, and taste. In this sense they are just like the living, except for one crucial detail: they cannot touch and they cannot be touched. Odysseus realizes this when, after seeing his mother, he tries in vain to embrace her not once, or twice, but three times; her shadow slips through his hands like a statue made of mist. Odysseus complains and laments. His mother gives him solace with these words: "This is just the way of mortals when we die. Sinews no longer bind the flesh and bones together."[15] Our sinews, masters of all balance and, thus, of proprioception, are the foundation of all living movement. Their haptic primacy is the cornerstone of experience in the living body.

Life is thus defined by the ability to touch, to be touched, and to feel one's body as a balanced organism that can move. One of the ghosts summoned by Odysseus is Achilles. His opaque specter laments the fate that was assigned to him and confesses that he would rather be a slave among the living than a king among the dead. After Odysseus tells him that his son, Neoptolemus, has become a fierce and successful warrior, Achilles turns around and moves away slowly, somber and intangible, through an asphodel field. When he was alive, the great hero of Troy had also learned that death is an endless state of intangibility. In book 23 of the *Iliad,* Achilles is being consumed by grief over the death of Patroclus. In spite of having already killed Hector and defiled his corpse, he still does not feel an ounce of solace. Sad to the point of exhaustion, Achilles eventually falls asleep and the ghost of his friend visits him in a dream to implore him to check his rage and, once and for all, celebrate his funeral. Achilles agrees and begs Patroclus to embrace him: "he stretched his loving arms but could not seize him, no, the ghost slipped underground like a wisp of smoke with a high thin cry. And Achilles sprang up with a start and staring wide, drove his fists together and cried in desolation: 'Ah god! So even in Death's strong house there is something left, a ghost (*psyche*), a phantom (*eidolon*), true, but no real breath of life (*phrenes*)."[16]

Homer uses the term *psyche,* normally translated as "breath," or "spirit," almost exclusively to refer to the dead. Similarly, *eidolon*—"image," or "phantom"—is mostly used in reference to ghosts. To describe the interior life of the living, the most common terms are *thumos, nous,* and *phren*; the first one commonly applies to the soul as origin of emotions and basic desires, and especially as the seat of feelings like anger and affection. The second—translated commonly as "mind"—has to do with perception and intention. And the third refers to both emotion and rational thought. *Phren,* which can be translated as "mind," "breast," and "spirit," is sometimes identified with the diaphragm, although some claim that for Homer, who almost always uses it in the plural, it refers to the lungs.[17] As the seat of thought, but also of emotion, *phren* is a principle of movement—movement that leads from feeling to action. Unable to hold the ghost of Patroclus, Achilles understands that the dead are *psyche* but have no *phren,* thus confirming the haptic nature of the latter faculty. This is an epiphany for the hero and he claps his hands in an outburst of surprise—a wonderful eureka moment in the midst of the anguish of mourning. Two other terms refer tangentially to *phren* and, more directly, to the heart: *kradie* ("heart") and *etor*, another word for "heart." In the convoluted and asystematic Homeric psychology, the *etor* is commonly located in the *phren*. *Etor* is a source of emotions that, after having passed through the complex refinery of the *phren*, provoke reactions and bring forth decisions. The *etor*, understood as heart, or rather as "vein," is thus a primordial beat, the primal impulse that originates feelings such as anger, which constitute the very haptic core of the *Iliad*.

I suggested before that the *Iliad* was a succession of affective outbursts in a wasteland of narrative dilations, standstills, and dramatic tension. But the *Iliad* is also a journey through Achilles' complex interiority. Whereas Odysseus in his *nostos* faces monsters, witches, sirens, lotus-eaters, storms, and the wrath of the gods, Achilles' own odyssey is beset by just one obstacle that, growing from the depths of his own self, proves to be considerably more challenging and potentially much more dangerous: his rage. The first word of the poem is the noun *menin* (accusative form of *menis*, meaning "rage" or "rancor"), modified in the second line by the adjective *oulomenen* (murderous). "Rage, goddess, sing the rage of Peleus's son Achilles, murderous . . ." Homer also uses the term *cholos* ("wrath"). Achilles' rage and wrath burst like volcanic eruptions from the innermost depths of his interiority, from his *thumos* and his *etor*; they drive the plot and run through the poem like wildfire. Twenty-four books and 15.693 lines later, the *Iliad* concludes with the word *hippodamoio* ("horse breaker"), the genitive form of an epithet used to characterize Hector. *Damoio* comes from the verb "to break," or "to tame." This reference to the taming of wild horses—an ancestral symbol of the passions, which Plato would later immortalize with his chariot simile in the *Phaedrus*—closes a poem whose main theme is the devastation caused by unbridled emotion. In a memorable scene from book 24, Achilles proves he has learned to check his own rage. Faced by Priam, who has made his way to the hero's tent to beg for the body of his son, an observant and self-aware

Achilles anticipates a tense situation that might lead him to a choleric outburst and avoids it. It is a remarkable moment of foresight, compassion, and restraint:

> Then Achilles called the serving-women out: "Bathe and anoint the body, bear it aside first. Priam must not see his son." He feared that, overwhelmed by the sight of Hector, wild with grief, Priam might let his anger flare and Achilles might fly into fresh rage himself, cut the old man down and break the laws of Zeus.[18]

Back in book 1, as Achilles was about to slay Agamemnon, Athena came down from heaven, held him by the hair, and controlled him, as one controls a wild horse. Now, however, there is no longer need of divine intervention; Achilles can restrain himself. What holds him this time is a self-inflicted squeeze that prevents anger from flowing through his throbbing *etor*—the invisible hand of self-control.

In more ways than one the *Iliad* represents a haptic journey. Its most crucial conflicts unfold inside, in the dark realm of affect, where nothing is visible, where everything is tangible, amidst the emotional outbursts of the *thumos* and the beating of the *etor*. Homer enters and exits through the skillful tapping of his dactyls; he comes near and then flies away like the Olympian gods, moving faster than a free-falling anvil. He illustrates the fatal consequences of these microscopic, internal spasms by depicting the horror of war with unparalleled perspicuity and with unrivaled humanity. From rage—annihilator of life, destroyer of world—to self-control—seed of all civility—there is just a mere voluntary movement of pressure, a contraction, a systole. Yet Homer also understands artistic production as an endeavor articulated by the carefully orchestrated oscillation between distance and proximity, from touch to sight and back to touch. The haptic nature of art itself becomes apparent as we sink into Hephaestus's workshop only to come out immediately after, elevated by the creation of a new world in relief. This metallic and three-dimensional world of varied and well-carved textures is the world of the poem; it is the *Iliad* as a whole, but it is also our own world. Homer brings it so close to us that we can almost touch it. And we do.

TORN TO PIECES

In the winter of 1905, in Beijing, a Mongol prince, ruler of the Aohan Banner, was hacked to death with a meat cleaver. Arrested, expeditiously convicted, and sentenced for the crime was one of the victim's guards, a man named Fuzhuli. The punishment imposed in the Qing Code for crimes of such gravity (magnicides, parricides, matricides, and other "enormicides") was the infamous *lingchi*. Commonly translated as "death by a thousand cuts," *lingchi* consisted in tying the convicts to a post and slicing pieces of their flesh until they died. On a late winter's day, at Beijing's vegetable market, the executioner began cutting off large portions from Fuzhuli's chest, biceps, and thighs. Finally, he quartered and beheaded him. Once he finished, the executioner uttered the accustomed pronouncement: *Sha ren le* ("This person has been executed").

Lingchi was not meant to be a prolonged torture; the whole process lasted a few minutes and it was not uncommon for the executioner to stab the victim in the heart after a few cuts. Disproving what many scandalized Europeans who witnessed it reported, the cuts were not a thousand; they hardly amounted to about fifty. And often the convict was provided with opium to palliate the pain. Fuzhuli's execution was photographed and the pictures, published in Louis Carpeaux's *Pékin qui s'en va* (1913), circulated throughout Europe, and decades later were the object of Georges Bataille's aestheticist delight. Less than ten years after Fuzhuli was executed, China abolished *lingchi.*

Contrary to what Norwegian, English, French, and Spanish eyewitnesses to *lingchi* chose to believe, the purpose of this form of punishment was not to inflict inhumane pain, but to tear to pieces. *Lingchi* was reserved for the most abhorrent crimes because its goal was to undo what Sinologist and legal historian Melissa Macauley calls "somatic integrity."[1] The human body is an assembly of parts, limbs, organs, muscles, tendons, etcetera, that form an organic unity. The ability to perceive it as a unity is possible thanks to intraception, proprioception, and affect, some of the many faculties of the haptic. This overall perception of us as one is guaranteed by a primordial and underlying intuitive conviction of being an indivisible, integral whole. *Lingchi* is an attack against the credence of this very experience, because it reveals that, far from being a self-evident truth, it is a mere belief, an act of faith. The process of cutting and tearing that turns a human being into a pile of flesh reveals the true nature of the human body, which is divisible, fragile, and contingent. The opium provided to the victim is an interesting detail. Protected by the narcotic and analgesic effects of the poppy, Fuzhuli doesn't feel: he is insensitive to the ordeal. Even before the executioner makes the first cut, Fuzhuli is no longer a sentient body, but a heap of flesh waiting to be fractioned. As it stands tied to a post and undergoes in a trance the unspeakable process, the body of Fuzhuli is reduced to a didactic spectacle, not unlike a corpse in an anatomy theater. More importantly, the show is not for him, but for the others, the onlookers. Whoever witnesses the execution and sees how little it takes to turn a person into a pile of fragments goes home with a brutal *memento mori*. But they also leave having learned a valuable lesson about the true laws that rule human life: the laws of physics.

I first read about Fuzhuli's martyrdom during a break from the research I was doing at the Bibliothèque Nationale, in Paris. I had been awarded a fellowship to spend a month looking at early modern editions of Lucretius's *On the Nature of Things* and, often, as I waited for the books to be delivered, I browsed through the stacks of the basement. That is how I came across Louis Carpeaux's book and the grisly picture of Fuzhuli. My topic of research was the reception of Lucretius in Renaissance Italy and, in particular, in the work of one of the pioneers of epidemiology, Girolamo Fracastoro, who lived in the first half of the sixteenth century.[2] As it often happens when one is fully immersed in research, a strong monothematic (or, rather, monomaniacal) tendency drove me to connect the account of the appalling execution with the rediscovery of the Latin poet in fifteenth-century Italy. The more I thought about Fuzhuli's torment, the better I understood the fascination of early modern intellectuals with Lucretius. But I should start from the beginning, lest this association runs aground in the sands of the absurd.

It was the summer of 1549 and Giambattista Bussini, one of the many humanist clerics that Paul III's Rome swarmed with at the time, was witness to an extremely peculiar event. Perhaps I would be exaggerating if I said that the secular character of modernity was sealed in that very moment, but what took place during the meeting that Bussini describes to his friend, art critic Benedetto Varchi, is a clear picture of what the future of Western culture would look like; a future that the Church, in whose most intimate core the meeting took place, could not foresee and, faced with the prospect of which, would have been up in arms. During this particular meeting many issues were discussed, among which were some potential new inclusions in the catalog of forbidden books that, a decade later, would be institutionalized under the name of *Index librorum prohibitorum*. Four years into the Council of Trento, and under the growing threat of Protestantism, tensions in the Catholic Church were running high, and the list of books banned for being impious, heretic, obscene, Lutheran, anti-papist, filo-pagan, or just intellectually dubious was growing by the minute. On that occasion, someone suggested adding *On the Nature of Things* to the list, a Latin philosophical poem in dactylic hexameter composed during the time of the Late Republic by one Titus Lucretius Caro. There was no lack of reasons for such a suggestion. Lucretius was an Epicurean who believed that the world—the *universe* would be more appropriate—was eternal and infinite, and that the human soul was material, divisible, and, therefore, mortal. As for the divine (Lucretius was a polytheist), its agency was not of much interest to him, and he claimed that the gods were material entities composed of such subtle atoms that there was no way that their actions could affect us. He also believed that divine intervention, providence, and the idea of an afterlife of punishment, or rewards, were lies concocted by priests and theologians to keep the masses fearful, docile, and submissive to power. Lucretius's poetry is stunningly beautiful and strangely moving; his philosophy is convincing. *On the Nature of Things* could possibly be the most compelling antireligious work ever written. The poem had already been banned from schools in Florence in 1517; in fact, in the sixteenth century the term "Epicurean" was a synonym for "atheist." And yet, when someone during that meeting in the summer of 1549 suggested that Lucretius be banned, incredibly, miraculously, ridiculously, the proposal was not well received. Cardinal Marcello Cervini argued that there was no need for that since that type of pagan mythology was utterly harmless.[3] In the coming decades and centuries, Lucretius's poem, materialism, and atomism in general would go on to influence and inspire Giordano Bruno, Francis Bacon, Galileo Galilei, Pierre Gassendi, Thomas Hobbes, Isaac Newton, Denis Diderot, Charles Darwin, Karl Marx, Albert Einstein, Erwin Schrödinger, and many others. Materialism and the notion that nature alone holds the keys to unravel its own secrets are the epistemological pillars of atomistic thought, and were instrumental to the staggering progress of early modern science. Although Cervini's pragmatic stance did not help to promote Lucretius—by then the

Latin poet was already a widely read and admired author in Italy—it can be read as an ominous sign that the secular road that modern thought was slowly taking was inexorable. Finally, this secular turn goes hand in hand with a progressive revaluation of the role of touch, the sense that Lucretius considered to be the cornerstone of all existence and of all perception.

Book 2 of *On the Nature of Things* opens with a man on the top of a cliff watching a storm unleash over the sea: "Pleasant it is when on the great sea the winds trouble the waters, to gaze from shore upon another's great tribulation."[4] The poet reflects on the strange combination of fear and self-assurance that someone observing a catastrophe from afar often feels. The pleasure that one experiences is not *Schadenfreude,* but mere relief, a consequence of being temporarily safe from danger. Similarly, whoever manages to rise up in knowledge thanks to philosophy and understands the nature of things shall be rid of the two most crippling fears that condition our lives: the fear of death and the fear of the gods. This distancing oneself, this intellectual detachment, produces a type of pleasure that Lucretius characterizes throughout book 2 with the tactile adjective *suavis* ("soft"). The Epicureans famously held as the highest good the pleasure of avoiding pain through wisdom and detachment, here described as a softness that caresses the soul—which is also made of atoms. As a way of introducing Memnius—the young addressee of the poem—to the benefits of Epicurean philosophy, Lucretius dedicates this second book to unraveling the complex atomistic ontology and epistemology. Both center on an idiosyncratic notion of touch that resonates remarkably with the concept of the haptic.

Lucretius wrote *On the Nature of Things* toward the middle of the first century BC. We know that the poem was well received by some of his contemporaries, such as Cicero, and soon after by other poets such as Virgil and Horace. From the early Christian years on, though, and through the Middle Ages, the work was almost completely forgotten. In 1417 a humanist adventurer, book hunter, and explorer of remote libraries, Poggio Bracciolini, found the manuscript in a German monastery and immediately knew that he'd come upon something in which his friends in Italy would be interested. He was right. Less than a century later, Lucretius was already part of the classical canon and his revolutionary ideas were already surreptitiously transforming Western thought.[5] Two of the greatest poets of the fifteenth century, Angelo Poliziano and Giovanni Pontano, revered him. Pontano said: "[Lucretius] takes his readers wherever he wants to go."[6] Philosopher Marsilio Ficino, editor and printer Aldo Manuzio, and Machiavelli were some of the luminaries that fell under the spell of the Latin poet. The poetry enamored them, but the ideas shook them to the core. As a young man, pious Ficino became enraptured with the reading of *On the Nature of Things* and even wrote a commentary that he later burnt—I "gave [it] to Vulcan," he confesses to a friend both ashamed and relieved.[7] Doctors must sometimes sweeten the rim of the cup that contains the bitter medicine to encourage the patient to drink it and heal, says Lucretius. His poetry is the honey that hides the sharp flavor of knowledge. The verb that Lucretius uses to refer to this smearing of the rim is *contingo*, a composite of *tangere* ("to touch") and the preposition *cum*, from which the word "contagion" comes. Contagion is a type of touch that comes with something, that leaves something behind, a mark, a stain; it is a transformative touch. Lucretius's grand entrance in the intellectual scene of fifteenth-century Italy is also a contagion. His poetry and his philosophy—inseparable one from the other—penetrate the bloodstream of European humanism like an incurable disease. Europe catches Lucretius and becomes infected with his haptic poetry and his philosophy of touch, as one usually catches a virus: inadvertently.

According to Lucretius, who claims to be divulging the philosophy of Epicurus, everything that is, is either body or void. In book 3, the poet says: "The nature of mind and spirit is bodily; for when it is seen to drive forward the limbs, to arouse the body from sleep, to change the countenance, to guide and steer the whole man, and we see that none of these things can be done without touch, and further that there is no touch without body, must we not confess that mind and spirit have a bodily nature?"[8] To be a body means to be able to touch and be touched; bodies are tactile and tangible. Unlike what many believed, the atomists were not atheists. They indeed believed in and reverenced the gods, but they argued that, like everything else, the gods too are material bodies consisting of atoms—atoms so subtle that they cannot affect us or the world that we live in. However, even though the atomists believed in the divine, they rejected teleology and the idea of divine providence. Nature is the product of an ever-ongoing process of trial and error; there is no master plan, or fate, or intelligent design. As modernity comes into contact with materialism, the teleological conception of nature proposed and promoted, in a variety of forms, since Plato and Aristotle, through Galen, Saint Augustine, Aquinas, and others, slowly but surely begins to wane. Modern science is ultimately Epicurean because it condemns the notion of final cause to obsolescence, and because it leaves no room for superstitions like afterlife punishments and eternal salvation. The only reality that matters is that of bodies, and the body works through a combination of haptic operations.

"For touch (*tactus*), so help me the holy power of the gods, it is touch that is the bodily sense (*sensus*)!" proclaims Lucretius in book 2.[9] By *sensus* he means sensibility in general; as for *tactus*, the variety of operations that the word refers to in Lucretius greatly exceeds the meaning of the English word "touch." In the context of *On the Nature of Things*, haptic would therefore be a more appropriate translation of the term. To feel, to sense, means to touch, and *sensus* is our only gateway to the outside world. We sense with different degrees of intensity even though some sensations are so subtle they go unnoticed. Everything touches us, we touch everything, and nothing is untouchable, not even the soul, not even the gods. But the haptic is not just at the basis of perception; it is also the foundation of Lucretian ontology and physiology. Atoms, which are eternal, imperturbable, impermeable realities, fall through the void and, as they collide with one another, they group and form composite bodies; thus all bodies are conglomerates of innumerable atoms in constant movement and in perennial contact. If we were able to see the world through atomic (or Lucretian) goggles, we would see swarms of atoms buzzing and vibrating, coming together and disaggregating. Some would look like sea urchins and Bengal tigers, others like daffodils, or pearls, sharks, birch trees, rabbits, emeralds, copper, a shooting star, men and women, an eagle, a narwhal, etcetera. Unlike the atoms that compose them, bodies are corruptible and perishable. It is precisely that constant, buzzing movement of their elements that progressively consumes them, until one day, with the body either dead or completely eaten away by atomic erosion, the swarm disperses and flies away to produce new bodies.

In an essay on ancient materialism, Erwin Schrödinger observes that the atomism of Democritus and Lucretius and twentieth-century atomism are merely different stages in the evolution of one theory. For Schrödinger, the same core idea informs both positions, namely an attempt by human imagination to "bridge the gulf between the real bodies of physics and the idealized geometrical shapes of pure mathematics."[10] Ever since the days of Democritus and Lucretius, all the way to Einstein through Galileo, Descartes, Marx, and others, says Schrödinger, the task of the atomist has been to facilitate "our thinking about palpable bodies."[11] By watching a body disintegrate (as in those time lapses that show a rotting carcass, or in the footage of the demolition of a building), we grasp with shocking vivacity both the essential palpability of the real and its intrinsic divisibility. So too, the photographic testimony of Fuzhuli's execution by *lingchi* exposes the human body for what it is, a summation of parts, and illustrates, in the crudest and starkest manner, the overwhelming fact that everything is matter and that all matter is divisible.

But let's go back to our story—the story that Stephen Greenblatt's *The Swerve* tells with such evocative elegance. It is the summer of 1417 and Poggio Bracciolini is going through ancient manuscripts in the library of a German monastery. A particular codex catches his attention, so he sets it aside and dusts it. It was almost certainly daytime. Libraries closed after dark due to the danger that candles posed to the precious collections. With the volume in his hands, perhaps the Tuscan book hunter blows and centuries of dust scatter all around forming a cloud; the tiny, almost imperceptible motes fly off and float lightly in the diaphanous rays of sunlight that come through the windows. The book is, of course, *On the Nature of Things*, and the image of dust motes floating in a sunbeam is one of Lucretius's favorites to illustrate the dance of atoms in the void. So, as Poggio begins to browse through the text and realizes that he has found something special, the dust disperses and is blown away; it goes out of the window and flies south, away from the monastery in the northern wind, following the course of a river maybe, as if attracted by a magnetic force. We know where it is going. Italy is its magnet, its long-lost motherland. It flies over Ulm and Innsbruck, Bolzano and Trento, it rushes through Mantua—Virgil's homeland—and gets to Florence first, then Rome, finally Naples. In its vertiginous flyby, it brushes against mountains and meadows, towns and villages, goats and goatherds, flowers, springs, cliffs, and caves. All of this and much more is what Lucretius's poetry is made of—it is a poetry of the senses, and in the case of Lucretius this means that it is a poetry of *tactus*—a haptic poetry. When he sees goats and sheep with bustling udders, or the shiny emerald green of the valleys, when he smells fresh milk and wild flowers that someone has picked to make a crown, when he drinks the virginal water from a mountain spring and listens to the crystalline sound of it flowing, Lucretius is neither seeing, nor smelling, nor tasting, nor hearing: he is touching and being touched. In his words:

> Lastly, the raindrops pass away when father ether has cast them into the lap of mother earth; but bright crops arise, the branches upon the trees grow green, the trees also grow and become heavy with fruit; hence comes nourishment again for our kind and for the wild beasts; hence we behold happy cities blooming with children and leafy woods all one song with the young birds; hence flocks and herds, weary with their fat, lay their bodies about the rich pastures, and the white milky stream flows from their swollen udders; hence the young ones gambol in merry play over the delicate grass on their weakly limbs, their tender hearts intoxicated with neat milk. Therefore no visible object utterly passes away, since nature makes up again one thing from another, and does not permit anything to be born unless aided by another's death.[12]

The original Latin passage is dominated by strikingly sonorous alliterations of f, r, and, especially, p sounds. The combination of the two voiceless consonants (the occlusive p and the fricative f)—pure impact, no vibration—and the alveolar trill of the r, create an effect of erupting linguistic exuberance, at once sizzling and explosive, that reproduces the scene of bucolic abundance, verdant fertility, and voracious lactancy that the images evoke. The sound of the words precedes the sense and, to a certain extent, prevails over it, determines it even. Here, as in many other passages in the poem, Lucretius's dactylic hexameters tap into our innermost fibers like sprightly fingers following the rhythm of music on a hard surface. The result is a powerful kind of song made of pre- (or post-) linguistic affect and of the music of words; something akin to what Robert Frost used to call "the sound of sense."

The instant fascination with Lucretius that some of the most prominent poets, intellectuals, and natural philosophers in the Renaissance felt was, in part, the product of a strong affinity with the poet's tactile hypersensitivity; an intuitive sort of understanding, perhaps. But understanding implies some degree of mediation and detachment. Perhaps it is more accurate to say that the Renaissance grasps Lucretius. It *gets* it and it is, in turn, captured and caught by its beauty and its poignancy. For this aesthetic symbiosis to happen, reciprocity is crucial. The Renaissance is receptive to Lucretius's exacerbated tactile sensibility because its own sensibility is changing and becoming more tactile. The rediscovery of the Latin poet coincides with a revaluation of the lower senses, especially touch, after over a millennium of their being denigrated and neglected by the most prominent thinkers in the West. But, in addition to this turn to tactility, there is another factor that conditioned and shaped the reception of Lucretius in Europe: the effects of the two greatest pandemics of the sixteenth century, syphilis and smallpox. Syphilis was a novelty at the time. It most likely arrived in Europe from the new world as Alonso Pinzón's *La Pinta* docked in Baiona, a small port in Galicia, in March of 1493. Smallpox, on the other hand, was well known to Europeans, but throughout the sixteenth and seventeenth centuries outbreaks become more frequent and the scope of their devastation augments exponentially, in a large measure due to the rapid growth and the concomitant overcrowding of urban centers. These two ailments wreak havoc in the body and cause a level of physical destruction of the likes and scale not known until then. One of the first accounts of syphilis comes to us from physician Ruiz Díaz de la Isla, who describes with hair-raising precision the chancres, ulcers, and craters that consumed Alonso Pinzón's body. The D-day for the first outbreak of syphilis in Europe is February 22, 1495. At approximately four in the afternoon, the troops of Charles VIII of France—composed mainly of Spanish mercenaries, many of whom were syphilitic—took the city of Naples. Once the assault was over and the troops disbanded, like a swarm of atoms, men and women spread the disease throughout the continent, reaching all the way to North Africa and the Ottoman Empire. The effects of the bacterium *treponema pallidum* on the body are ghastly and physicians of the time describe them with a combination of horror and morbid curiosity. This is what Sicilian doctor Niccolò Squillaci says after seeing a number of patients in 1495:

> The purulent pustules spread in a circle, and there is an abundance of the most virulent lupus. The signs of the sickness are these: there are itching sensations, and an unpleasant pain in the joints; there is a rapidly increasing fever, the skin is inflamed with revolting scabs and is completely covered with swellings and tubercules which are initially of a livid red color, and then become blacker. After a few days a sanguine humor oozes out; this is followed by excrescences which look like tiny sponges which have been squeezed dry; the sickness does not last more than a year although the skin remains covered in scars which show the areas it affected. It most often begins with the private parts. . . . I exhort you to provide some new remedy to remove this plague from the Italian people. Nothing could be more serious than this curse, this barbarian poison.[13]

Three decades later, Girolamo Fracastoro coins the term "syphilis" and writes a poem in dactylic hexameter inspired by Lucretius, where he explores the symptoms, causes, and cures for the disease. Fracastoro's descriptions are both beautiful and atrocious. First, the ailment pollutes the blood, making it dense, filthy, and viscous. Then it explodes like a range of volcanoes on the surface of the body:

> At once disfiguring sores broke out over the entire body, basely defiling the deformed face and the chest. Then the disease took on the new appearance of a pustule that, resembling the top of an acorn, was rotten with thick pus. Not long after, it burst open, unleashing a multitude of phlegmlike blood and disease. Soon the affliction gnawed into the body and, hiding deep therein, mercilessly consumed it. Often I myself have seen limbs and ravaged bones shorn of their flesh, while the mouth and throat, gnawed away and gaping in a sordid wound, could emit only the feeblest sound. It resembled the thick fluid that one often sees flowing from the thick bark of the cherry tree or the sad almond tree of Phyllis and that quickly hardens into a thick gum. Very similar indeed was the fluid that this disease spread over the afflicted body and that then calcified into filthy scabs.[14]

From the late fifteenth century until the invention of penicillin and the first successful treatments in the 1940s, syphilis disfigured, maimed, and killed hundreds of thousands of people.

The Americas gave Europe syphilis; Europe, in turn, reciprocated with smallpox and the flu. In the sixteenth and seventeenth centuries the *variola* virus decimated populations on both sides of the Atlantic with never-before-seen ferocity. It would not be far-fetched to speculate that, during that time, especially in larger urban centers, those who had not suffered (and survived) the disease had at least been exposed to it by witnessing its effects on others. And what this virus does to the human body is simply horrifying. In his chronicles of the conquest of Mexico, Franciscan friar Bernardino de Sahagún describes the first outbreak of smallpox among the Aztecs in the fall of 1520. Their bodies were covered head to toe with pustules, and the sick died of hunger and thirst because nobody would dare to come close to them. Those who did not die remained scarred for life, their faces disfigured by gashes and craters, their skin permanently lacerated, as if their bodies had been carved with the sickle of the angel of death. So appalling was the deformity that many survivors went into reclusion or killed themselves. The damage was at times so severe that people's physiognomies were utterly effaced and their very identities lost. In 1884, during an outbreak in London, nurse Isla Stewart tells the story of two survivors, two sisters, who ate together at the same table in the hospital for six weeks before they finally recognized each other.[15]

The periodic orgies of syphilis and smallpox put the body on the forefront, exposing it as a fragile, shameful, pestilent, divisible, and contagious entity. While in the streets men, women, and children suffered the scourge of the plague, in the anatomical theaters physicians were opening cadavers ever more regularly and discovering the mysterious and hitherto virtually unknown inner world of the human body. Toward the late Middle Ages anatomists had begun dissecting corpses, and by the seventeenth century the practice was standard procedure in most of the major medical schools around Europe. Many in the Renaissance were receptive to the Lucretian concern with matter and bodies because they themselves, plagued by epidemics and exposed to the insides of the body, were invested in unraveling its mysteries. And one thing that anatomical dissections and diseases like syphilis and smallpox expose is the intrinsically divisible nature of the body, its frailty and contingency, its irrevocable materiality. It is in the face of this revelation that the sense of touch—considered for centuries unworthy of epistemological attention—acquires epistemic, but also aesthetic and even moral, relevance. Anatomists like Alessandro Benedetti and Andreas Vesalius realize that, in order to know the body, one needs eyes and hands to work in tandem. Epidemiologists like Fracastoro understand that diseases are transmitted through physical contact between an infected body and a healthy host, and not by magic spells, or as a result of divine punishment. Fracastoro, influenced by Lucretian terminology, suggests that contagion is precisely, as the word says, a "touch with," a type of touch that carries with it an added bonus, that is, infection.

But not all contagion is lethal. Lucretius conceived his own poetic and philosophical mission as a form of contagion and bodily invasion. The poet-philosopher does not appeal to mere persuasion or aesthetic seduction; he inoculates his concepts into the listener in a physical, or rather corporeal, attempt to vanquish pre-established and incorrect beliefs that also inhabit the material soul. Human beings have been persuaded that they have immortal souls, that they will be punished or rewarded in the afterlife, and that there is a force of providence, or destiny, that rules the world, because they are terrified by uncertainty and by the prospect of complete annihilation. In the opening of book 4, Lucretius clarifies: "My teaching is of high matters and I proceed to set free the mind from the close knots of superstition; next because the subject is so dark and the verses I write so clear, touching (*contingo*) every part with the muses' grace."[16] *Contingo* is a composite of *tango* ("to touch") and can mean "to soak," "to impregnate," "to lather," but also "to infect." For Lucretius, the mechanism of poetry—like that of knowledge, like those that articulate generation and corruption—is corporeal, and that means that it is tactile, or, better, haptic. The poet is touched by sensations that inspire verses that, in turn, touch whoever listens or reads them. These verses—melodious, mellifluous, at times sublime—also carry with them teachings that operate in us tangible changes in the way we understand the nature of things. Additionally, the truths of Epicureanism, Lucretius hopes, are absorbed with great pleasure. What is death? Lucretius asks and subsequently invites us to ponder what we were ten months before we were born. The answer is simple: nothing. And that is precisely what death is, a state of absolute nothingness.[17] The atoms that compose us disperse and go back to floating in the void, until the vicissitudes of their own gravity, the random and uneven movement as they fall (Lucretius calls it *clinamen*, a "swerve"), group them with other atoms and they compose something else, an otter, or a diamond, an asteroid, or a coral reef. Many astrophysicists today agree with the broad assumptions of such a view and even claim that these atoms that we are made of, that everything is made of, are the remnants of stars that exploded billions of years ago. Indeed, they add, the atoms of your left hand most likely come from the dust of a different star than those of your right hand.

How is it then that bodies deteriorate? How is it that invisible, microscopic beings have the power to eat away flesh and bone, to erode rocks, to corrode metal? Doesn't the air move the trees, sometimes violently enough that it brings them down, without us being able to see it? asks Lucretius. Don't rings wear away in the fingers of spouses as the decades go by? Or the statues of the gods, are they not gradually consumed by pilgrims kissing them year after year, century after century? Sight does not allow us to perceive this because "nature works by means of bodies unseen," by means of blind bodies that touch.[18] We cannot see it, we often cannot feel it but our bodies are wearing away, abraded by the constant rubbing against other bodies and by the buzzing of the atoms that constitute us. The poetry of Lucretius enters through eyes and ears, it reaches the soul and collides against it, it impregnates the intellect and infects it with new doctrines that defy tradition and superstition. Pontano said it best: Lucretius takes you wherever he wants to go. The Latin poet holds, catches, shakes, moves, and drags his reader at will. The tactile language is not metaphoric. Lucretius is a materialist; when he says that he touches everything, when he says that he unshackles the mind, he means it literally.

Perhaps one of the moments that better illustrates the complex implications of the notion of *tactus* is the final section of book 4, where Lucretius addresses the causes of sexual desire.

The poet's bleak and naturalistic approach anticipates by almost two millennia Schopenhauer's *Metaphysics of the Love of the Sexes*. The male seed boils up in the lower abdomen and strives to burst out toward the object of desire, notes Lucretius. The lovers then intertwine hungry for one another's body, hoping to possess each other, or to become one. But satisfaction is denied to them and possessing the other is impossible. Lucretius explains: "They cling greedily close together . . . ; but all is vanity, for they can rub nothing off nor can they penetrate and be absorbed body in body, for this they seem sometimes to wish and to strive for, so eagerly do they cling in the couplings of Venus."[19] The despair of lovers is depicted almost in a humorous tone, but the reality of their conundrum is both tragic and animalistic. Their bodies crave for one another, they desire to become one—a reference to Aristophanes' bizarre myth in the *Symposium*, perhaps?—but the atoms that compose them (immutable, impermeable, indestructible) will not allow it. This is the reason why, for Lucretius, sexual desire is impossible to satisfy. Even in the most intimate moment of intercourse, there is nothing but superficial contact, a mere tingling. Sometime after the climax, desire comes back and the cycle recommences. Lucretius conceives sexual desire as an inconvenience, a distraction, and a source of anxiety and frustration that steers us away from the life of *ataraxia*, that state of focused calm free of strong emotions for which the Epicurean yearns. But sexuality is also revealing of a curious dissonance between the senses and, in this way, functional to further analysis of the haptic. When stung by the bee of desire, lovers do not know whether first to enjoy each other with their eyes or with their hands. If they come close enough to kiss, they cannot see each other. From a distance, they can contemplate each other's beauty, but they cannot touch. Even though for Lucretius sight *is* touch, it is a much subtler form of contact that hardly helps to quench the lover's lust. The yearning for proximity always prevails. But, alas, touch can never be more than superficial contact; and this confirms the materially discrete character of corporeality. The frontier zone where contact takes place, far from consolidating proximity, intensifies the sense of separation and division, because it corroborates the impossibility of transcending the limits of one's body. The haptic nature of sight then becomes evident.

Although for Lucretius all five senses are variants of *tactus*, the differences between them are noteworthy. Against a philosophical mainstream that since Plato had considered it to be the most noble and trustworthy among the senses, sight is of particular concern to the Epicurean poet because he deems it the most base and deceitful. From the top of a mountain we see down below in the valley a large white spot, like a cotton canopy, or a fallen cloud, immobile, homogenous. It is, in fact, a flock of sheep, dozens of animals each with their own distinctive, particular ovine features, slowly moving, chewing and ruminating, shaking their tails; the ewes have their udders bursting with milk that their lambs suck avidly, with such voracity that they cannot gulp it all and it streams down their young lips.[20] Only from up close can one appreciate the detail and the particularity of each individual; no two things in nature are exactly alike. Proximity is the realm of touch, and this means that only through it can we access the overwhelming variety of textures that compose the world. In spite of this, some of the most memorable images in *On the Nature of Things* are visual: the storm over the sea, the armies marching to battle like ants, the wild and cosmic dance of Venus *generatrix* that fifteen hundred years later inspired the poet Poliziano and his friend Botticelli. For Lucretius, the great champion of *tactus*, the goal of life is achieving tranquility, *ataraxia*, and this implies distance, detachment. What might seem like a paradox is, really, a matter of degree and distance. Sight too is touch, but of a much subtler kind. It is impossible to remain untouched, but the subtler and the more distant the touch, the less disruptive it will be. The model for this type of impassibility are, of course, the gods, made of such subtle atoms that circumstances cannot affect them in their distant, retired existences, from where they can grasp with the utmost immediacy the nature of things. So too, the philosopher standing on the top of the cliff watching the wind and the waves shake the fragile ship, remains undisturbed. Ultimately, what the philosopher sees from this position of imperturbable tranquility is not the vast expansion of reality, but the ocean of time. In the distance what the philosopher grasps through sight is the constant movement of atoms in the void, their gatherings and their separations that make bodies and undo them. One could picture this spectacle of reality unfolding as a never-ending succession of cycles of generation and corruption in the form of Borges's *Aleph*, a ubiquitous, simultaneous, and extemporaneous glimpse at the universe in all its excruciating variety as an all-encompassing time lapse. This spectacle is of course haptic, and it is accessible to the mind trained in philosophy; that of Lucretius is a particularly refined form of hypersensitivity.

Everything is always in a constant process of wasting away and becoming undone. This, when considered alongside the cruel dynamics of sexual desire, results in a world of yearning bodies that not only cannot fulfill their wish of becoming one with other bodies, but also cannot preserve their own organic integrity and progressively wear off. The intrinsic ambivalence of sexual desire (always bittersweet, never satisfying) demonstrates for Lucretius that the mechanism by which living beings reproduce is governed by the spiraling dialectics of desire and dissatisfaction. This, in turn, exposes the most basic paradox of existence: while we operate under the assumption that we are whole and homogenous beings, we are, in reality, but an assortment of parts stuck together by chance and habit. The irony could not be bitterer: the craving to penetrate other bodies and merge with them is met with the harsh reality that, between bodies, there is not more than superficial grazing. For

Lucretius, this reveals the most fundamental aspect of the nature of things: everything is made up of invisible, immutable bodies that come into contact with other such bodies forming larger, sensible entities and without ever changing their atomic shapes. Such are the lessons that the words of the poet carry, and they impact against our innermost core causing a stir that resembles a contagion rather than an intellectual epiphany. Beautiful language drenches our mind that absorbs the knowledge inadvertently, painlessly. For Lucretius poetic language, insofar as it is haptic, is effectively transformative; an attempt (perhaps hopeless) to harness into the unity of the poem through the artifices of language, prosody, and meter, an infinite disarray of bodies scattered by constant movement and corroded by permanent degradation.

During that stay in Paris I held in my hands several manuscript editions of Lucretius's poem, as well as incunabula, and sixteenth-century print editions. One afternoon, for a few hours, I had the chance to handle the *editio princeps*, published by Thomas Ferrandus in Brescia in 1473. At the end of that day of work, with the olfactory particles of the fifteenth century still tingling fresh in my nose, as I was leaving the library I saw a poster announcing an exhibit of the work of Joel-Peter Witkin. The next day I visited it, and as I stood in front of Witkin's photographs, what I saw took me back to Lucretius, to the execution of Fuzhuli, and to the fundamental paradox of corporeal and organic life. One picture in particular, a toned gelatin silver print titled "Still Life, Mexico" (1992; see fig. 2), from the collection *Twelve Photographs* (1993), caught my attention. Witkin is fascinated with human body parts, and during the 1990s he spent long periods of time in Mexico City working at hospitals and morgues. One day, as he was going through a box full of severed human limbs, wrapped in a cloud of cadaverous stench (there is no smell more powerful and abrasive than that of human decomposition, Witkin notes), he found half the

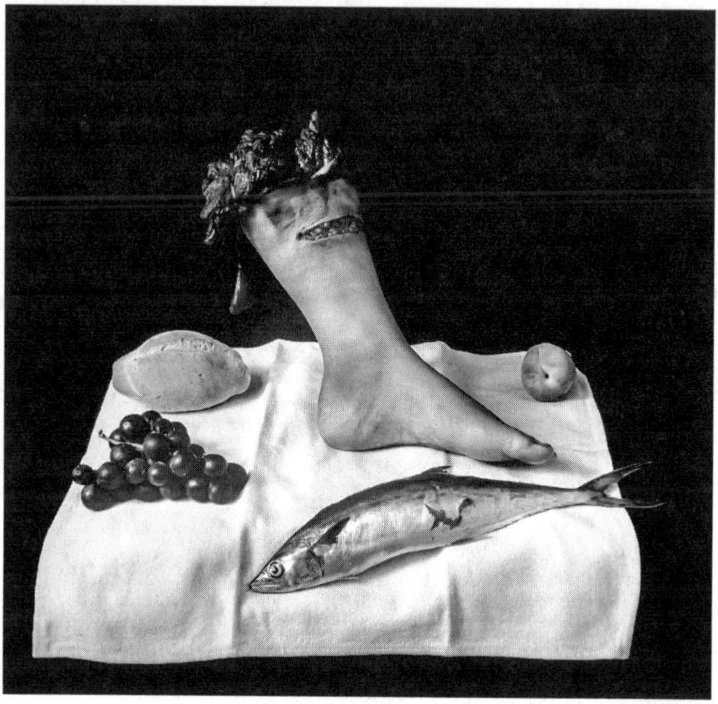

2. Joel-Peter Witkin, *Still Life, Mexico* (1993). © Joel-Peter Witkin. Courtesy of Etherton Gallery.

leg of a woman who had been run over by a train.[21] The victim had survived the accident, he was told, and she was still living somewhere in the city, a detail that the photographer found particularly inspiring. That evening he had dinner at an old German restaurant where the thick, white, traditional tablecloth gave him the idea of how to photograph the limb. He purchased one of the tablecloths from the owner, and the next morning he went grocery shopping and bought bread, fish, and fruit.

Against a background of absolute darkness, the half leg seems to float lightly over the tablecloth. This lightness makes the leg look ethereal, alive even. It was on its way somewhere and it has either stopped of its own volition, or something detained it; the amputated limb is not dead, it is still. Also, despite its being severed, it is an organic part of two larger dimensions: one visible, the other invisible. The visible one, the one of the tablecloth and the grapes, the bruised peach and the shiny putrescent whole fish, hovers in the dark background like the material world floated in the void for the ancient atomists. The invisible one is the body from which the leg was severed, but also the human body in general, imposing its presence through the conspicuousness of its absence.

In staging the stillness and the life of the severed body part among elements that have also been torn from their own unique environments, Witkin's photograph dramatizes the fictitious understanding of the body as a homogenous totality. And as it does this, the picture creates a new, artificial self-contained reality, by patching together the atomized parts, by turning them into organic elements of a new totality, and by reestablishing the fiction of wholeness—not unlike what Lucretius does with words. What the Epicurean poet and the artist seem to be conveying in a similar tone, at once powerful and elegant, is that the only possible way to produce something truly whole and truly organic is through artifice.

That afternoon, as I stood in front of Witkin's "Still Life, Mexico," the detail that captivated me the most was a gashing wound in the upper calf, right below the line of amputation, that exposed a dark, fleshy texture resembling raw minced meat. I later learned that this was Witkin's favorite part of the work; to him, though, the substance oozing from the open wound looks like caviar—a necrophagous delicacy in an otherwise frugal banquet. But the wound is also a crevice that summons tactility. This invitation to touch is our gateway to the third dimension opened by the picture: the dimension of depth. Through this blind, labyrinthine space made of dubious textures we can only orient ourselves by means of the haptic faculties. As we focus on the wound and access its black, meaty treasure, by activating a haptic-inflected vision, the picture betrays its deepest meaning. The appeal to touch and the strong affective feedback it arouses complete the artificial harmony by giving effective unity to the elements, but also reveal the nature of corporeal existence, which is essentially divisible and disjointed. Such insights are only accessible to a haptic sensibility.

When Joel-Peter Witkin was growing up in Brooklyn he had a neighbor who had been injured during World War II and had a beastly scar—a huge hole, rather—in his neck. Witkin remembers how fascinated he was by this wound to the point that, one day, he plucked up courage and asked the man if he could touch it. The veteran agreed and Witkin still looks back on the experience as one of the most formative moments of his childhood. Another equally illuminating moment, which has become perhaps his most famous anecdote, happened when he was six years old. One morning, as he walked to school with his mother and his twin brother, they were witnesses to a fatal car accident. Amidst the commotion, Witkin suddenly saw something that was rolling down the street in his direction. It was the severed head of a little girl. The photographer says that, moved by an uncontrollable urge to touch it, he leaned down and reached out, but his mother immediately stopped him and rushed him and his brother away from the scene. The spectacle of the rolling head, with its shocking combination of death and movement, left an indelible mark in the imagination of the child who, over four decades later, would reformulate the scene, turning it on its head by juxtaposing quietude and light vivacity in his *Still Life*. In both cases, the experience leads to an uncanny conclusion: everything is fragmentary and the membrane that separates the living from the dead is as good as indiscernible. Also in both cases sight calls for touch, perhaps because the haptic is essentially restorative as well as a source of lasting comfort.

Joel-Peter Witkin's photography combines the visual and the tactile with remarkable success. The initial approach to his work is, of course, visual, but the palpability of its textures, the contrast between diaphanous and opaque, light and heavy, and the tangible density of its elements immediately summon tactility. Such collaboration between the two senses leads to the visceral effect that Witkin's work is known for, a conflation of repugnance, compassion, beauty, and abjection that leaves a lasting mark in the viewer. What Witkin does with the camera, Lucretius, who ends *On the Nature of Things* with the gruesome account of a plague that befell Athens, accomplishes by means of poetic language. For Lucretius words, like everything else, are material entities and, therefore, composed of atoms. They touch and affect; and as they gather together in all manners of combinations thanks to the skill of the poet, they create worlds that compete in tangibility with the world that we inhabit. In the sixteenth and seventeenth centuries, as Western thought was inoculated with materialism and Christianity underwent its traumatic process of atomization, the word, written and spoken, became a particularly sensitive issue and many died because of what they said and wrote. Among them was one of Lucretius's most brilliant readers, Giordano Bruno. Recalcitrant, vitriolic, and unrepentantly heterodox, Bruno doubled down on Copernicus's thesis proposing that the universe was not just heliocentric but also infinite. These ideas, along with his dangerous dabbling in the doctrine of the transmigration of the soul and a set of acrimonious opinions about the Church, got him convicted of heresy and sentenced to death. Bruno's thought is impregnated through and through with the atomistic notion that matter is eternal and reducible to infinitesimal units. In 1591, not long before his final return to Italy after long years in exile, he published in Germany what would be his philosophical testament, three poems in Latin: *De minimo*, *De monade*, and *De innumerabilibus sive de inmenso*. The first deals with the basic elements that material reality is made of, the second with numbers and elementary geometrical figures, and the last one advances the idea of the infinitude of the worlds. All three works bear the mark of Epicurean thought and, especially, of Lucretius's epic poem. Whoever is willing to dispense with the myopic glasses of the specialist will discover in the "absolute minimum" discussed in the *De minimo* a reformulation of the ancient atom and a conceptual prefiguration of that of Newton, Einstein, and Schrödinger: a way of thinking about reality based on the notion that matter is ubiquitous and hegemonic, and a way of thinking about bodies as essentially divisible.

During his long and painful trial in Rome, on several occasions Bruno agreed to abjure many of his most controversial ideas. Exhausted and frustrated with the mind games played on him by the inquisitors during the grueling interrogations, he eventually changed his mind and refused to make any retractions. "I have nothing to repent of," he said to the officials of Clement VIII, sealing his fate. The death sentence ensued. On February 17, 1600, Bruno was burnt at the stake in Campo de' Fiori. The spectacle of a man burning to death is almost inconceivable to us twenty-first-century Westerners. I try to picture it nevertheless. Perhaps the first ashes that fly away (those from his hair?) float momentarily over the bonfire, like motes in a sunbeam. Then, amidst crackling sounds and small explosions that unfortunately don't do much to muffle the screams, comes the charring of the flesh. When it is all over, there is nothing but smut and ash, soot and bone fragments. Bruno is undone, turned into infinitesimal fragments, reduced to his absolute minimum. His material remains are cast into the Tiber where they begin a journey of many centuries through rivers, seas, and oceans—where, swallowed by fish, they wind up a meal on some table or other, or they evaporate and ascend to the sky as moisture that then rains upon the water, or upon a meadow, over a village, a town, a metropolis, where they finally become part of that future that his colossal imagination had already conceived.

Je célèbre des jeux paisibles
Qu'en vain on semble mépriser,
Les vrais biens des âmes sensibles,
Les doux mystères du baiser.
—Claude-Joseph Dorat[1]

ELEMENTS OF PHILEMATOLOGY

In July of 2011, *Rolling Stone* magazine published an interview with Larry David in which the comedian discusses his personal life with great candor. When asked about his first steps in the world of romantic love, David says that before he had ever gone out with a girl his biggest source of anxiety was not the prospect of sexual intercourse. By then he had already been exposed to pornography and understood—or at least believed he understood—the basic mechanics of coition. What truly baffled him, making him feel inadequate and insecure, was the thought of kissing somebody. No one ever shows you, let alone explains to you, what happens inside the mouths when two people are kissing, David concludes. In *Une vie divine* (2006), Philippe Sollers points out something similar: "Pornography insists with the organs in order to distract us from the true inner passion, the one that takes place from one mouth to the other."[2] But is it at all possible to talk about what happens in the moist and dark space formed by two mouths closed onto one another? What lips and tongues do when two people kiss, does it even fall within the range of things that can be described? Does language have the verbs, adjectives, and prepositions to account for the variety of actions, movements, sounds, sensations that occur during the romantic kiss? And if it does, can it provide guidelines? Can it teach how to kiss well? I imagine that the perplexity of the creator of *Seinfeld* resonates with many young people who are soon to give their first kiss.

With this in mind, I would like to explore the mysteries of the kiss and attempt to sketch out the elements of a possible philematology (from *philema,* the Greek word for "kiss"). To be sure, there are many kinds of kisses. I would argue, however, that the kiss of brotherly and filial love, the kiss of peace, Judas's kiss, the kiss of death, the kiss hello are qualitatively different from the erotic kiss. This is the one that interests me: the kiss given with an open mouth. Already praised in the Song of Songs, this is the most complex and sophisticated of all kisses, first because it is given and received simultaneously and, second, because it involves much more than mere superficial contact. Lips, tongue, saliva, breath, and even teeth play a role in the erotic kiss; but also hands and arms, the pelvis, the waist, the shoulders, all fundamental elements to the grip that accompanies and, to a large extent, makes possible the kiss. Finally, the titillation it produces directly affects the sexual organs. In sum, the erotic kiss is given with the entire body.

We know that the erotic kiss—the tongue kiss—albeit not a universal practice, has been around for millennia. Statuettes, bas-reliefs, pictography, frescoes, and mosaics found in places as varied as Hindu temples, a Neolithic cave in Palestine—the Ain Sakhri lovers, a phallic figurine made by Natufian artists—the ruins of a brothel in Pompeii, etcetera, prove that the erotic kiss has been associated with procreation since remote times. Influenced by classical Greek and Roman statues, most notably versions of the myth of Cupid and Psyche (two of the most legendary kissers), artists like Canova, Rodin, Brâncuși, and others explored the trope in sculpture. Hayez, Klimt, Magritte, Lichtenstein, among many, did so in painting. Hollywood made the romantic kiss one of its trademarks. And yet, in all these cases, the kiss itself, the turmoil that unfolds within the confines of the mouths, remains a mystery. Painters and sculptures can portray the lips and bodies of those who kiss, their ecstatic facial expression, their mouths half-open, or sealed shut, but the kiss in itself proves inaccessible. As for film, the traditional Hollywood kiss is invariably dry and, therefore, chaste; it is the mechanic repetition of an iconographic liturgy rather than an attempt to fully unravel the osculatory phenomenon. And although sexual mores have relaxed and today we get to see here and there on the screen tongues, moisture, and rapid mouth movement, this still does not get us much closer to the actual kiss. It is perhaps poetry that has given us the most successful and explicit approximations to the erotic kiss, and namely a specific tradition of amatory poetry that originates in the Hellenistic period, flourishes in Rome during the years of the late Republic and early Empire, and resurfaces with a vengeance in the Renaissance. During the sixteenth and well into the mid-seventeenth centuries, this trend grows in popularity while philosophers, physicians, and intellectuals in general begin thinking about love with renewed enthusiasm, and begin discussing, more and more unashamedly, the role played by the body in erotic matters. In fact, never before and never again in Western history was the kiss thought of and written about more than during the Renaissance. This artistic, poetic, and philosophical engagement with the romantic kiss between the late fifteenth and mid-seventeenth centuries is directly associated with two phenomena that are also in themselves intimately related: the revaluation of the sense of touch and the renewed fascination with the classical tradition of amatory poetry. It is in this context that poets and love philosophers set out to delve into the corporeal experience of eroticism and the haptic dimension of love.

The kiss is the second most intimate form of touch. Like sexual intercourse, it involves the exchange of bodily fluids. The skin of our lips is one of the most sensitive areas in our body, and the tongue, a complex and hypersensitive muscular organ, combines the senses of touch and taste. Thanks to its mobility and flexibility, the tongue of the lover is able to explore the beloved's body with exquisite precision, as well as receive tactile feedback that impacts both epidermically and affectively. Also, as in sexual intercourse, the kiss combines the faculties of all five senses. Touch operates as cutaneous, exteroceptive sensation through lips and tongue, hands that grip, arms that hold, noses that brush and press against each other, chin, cheeks, etcetera; but also as intraceptive and affective feedback, because during the erotic kiss the heart rate tends to increase, the sexual organs titillate, and the stomach reverberates and buzzes filled with proverbial butterflies. The reciprocity of the romantic kiss also illustrates with great brio the fact that tactility always implies tangibility: to touch is to be touched. The kiss engages too the sense of smell: the extreme proximity with the other gives us access to their scent, especially their breath. Leaving all discussion concerning the importance of pheromones aside, no one would argue that it is hard to enjoy a kiss if one doesn't also enjoy the smell of the other person. Hearing also plays a role, for the sonorous range of the kissing marks its rhythm, its intensity, and thus directly affects the overall enjoyment. Although it is not a particularly visual affair, sight is not fully dispensed with during the philematic feast. In the almost absolute proximity of the kiss, lovers exchange intermittent glances—sideway glances that reveal askew glimpses of the mouth or the cheeks, an eye shut wide, a portion of the face, a close-up of the nose. Last and certainly not least, there is taste. Many have pointed out the relation between kissing and eating. Some believe that the kiss is the ancient vestige of habits related to nourishment, an atavistic oral fixation.[3] Toward the end of this essay, I will come back to this connection between kissing and eating, and to the anthropophagous aspect of osculation.

To go back to Larry David's point, the impossibility of accurately portraying or explaining what happens when two people kiss might be related to the fact that the only way to truly know what kissing feels like, and how one must do it, seems to be, simply, by kissing. If this were the case, philematology would be a purely practical form of knowledge, inaccessible to all language and logic. And yet, just like mystics throughout history have struggled to put into words the essentially ineffable experience of divine union, some have also attempted to describe, analyze, anatomize, and taxonomize the phenomenon of the kiss.

I now jump from the twenty-first to the nineteenth century, from Los Angeles to Copenhagen, from Larry David to Søren Kierkegaard, and from the question about the kiss to a fleeting attempt to answer it. In one of the entries of *The Seducer's Diary*, Johannes (a wonderful hybrid between Hamlet and Don Juan) confesses that he has been gathering material to write a *Contribution to the Theory of the Kiss*. Johannes is only interested in the heterosexual kiss: "A perfect kiss requires that the agents be a girl and a man. A man-to-man kiss is in bad taste, or, worse yet, it tastes bad."[4] For Kierkegaard, the kiss on its own has no philosophical value. A stolen kiss, a kiss by mistake, a kiss between men, between women, between children is not really a kiss, but a mere instance of superficial contact between mouths. The only type of kiss that is relevant to the philosopher, he says, is the one that comes from erotic desire and that produces sensual pleasure. He then proposes a taxonomy of kisses based on their sound. Unfortunately, language lacks enough onomatopoeias to account for the overwhelming variety of kissing sounds. The author, nevertheless, provides a tentative classification: there's the smacking kiss and the whistling kiss, the slushy kiss, the explosive kiss, the booming kiss, the full kiss, the hollow kiss, and the kiss that sounds like calico.[5] Kisses can also be classified according to tactile and temporal parameters, Kierkegaard adds. There's the tangential kiss, the kiss in passing, the clinging kiss, the long kiss, the short kiss. In the temporal category, the author is especially interested in one major distinction: that between the first kiss and all others. "This has nothing to do with sound, touch, time in general; the first kiss is qualitatively different from all others, very few people think about this."[6]

The whimsical Johannes, however, seems more interested in the originality of the topic than in the topic itself, because not long after bringing it up he flies off to other things. Speaking in strictly philematological terms, his treatment of the issue is a tangential kiss that leaves us thirsty for more. But the committed philematologist must go deeper and farther; a tangential kiss is not enough. In order to fully engage with the issue one must French-kiss it, make out with it, suck face, tongue wrestle, swap spit, play tonsil hockey. The range of verbs and idioms that refer to the erotic kiss is evidence not only of the complexity and allure of the act itself, but also of its intrinsic opacity. When he proposes the topic of the kiss, Johannes points out how strange it is that no philosopher has ever addressed it. This could be simply because philosophers cannot kiss; if they knew how to kiss, they wouldn't have become philosophers in the first place, the seducer speculates wryly.

But Kierkegaard was wrong. A sign of the unique fascination with kissing in the Renaissance is that the period gave us the only philosophical work exclusively dedicated to the kiss. *Delfino, or On the Kiss* (ca. 1577), is a dialogue by Francesco Patrizi in which the two interlocutors explore the spiritual implications of the pleasure experienced in the kiss.[7] Patrizi, a Platonic philosopher born in Dalmatia who became famous for his vitriolic attacks against Aristotelianism, comes to the topic of the kiss moved by a very similar concern to that which would later inspire Kierkegaard's seducer. Philosophy, which has often dealt with love, has blatantly overlooked the kiss, he says. Patrizi believes that the softness of the romantic kiss is the key to unravel the mystery of the mediation between body and soul, because the kiss is the frontier between matter and spirit. Like Kierkegaard, Patrizi is only interested in the heterosexual kiss, and the inquiry that the two characters in the dialogue

(Delfino and Patrizi himself) carry out is strictly based upon experience. They begin by assuming that both of them have kissed before and from there proceed to dissect the osculatory phenomenon, beginning with its softness and sweetness and then proposing a taxonomy that aims to be exhaustive. *Delfino* is, thus, a descriptive and experiential work of philematology; a natural history of the kiss.

For Patrizi the romantic kiss can be given on six parts of the body and in four different ways; this gives us a total of twenty-four types of kisses. The parts of the body, in ascending order of softness, are: hands, chest, neck, cheeks, eyes, mouth. The four ways of kissing are: with dry lips, with moist lips, biting, and with tongue. The kiss on the mouth, which one "gives and receives at the same time,"[8] is for Patrizi a special case. Since it is an encounter between two mouths, the possibilities are immediately and exponentially multiplied. The kissers can alternate between biting, licking, and sucking each other's lips, and they can either take turns to use their tongues, or do it simultaneously. The highest point of "sweetness," or "softness," Patrizi concludes, is when the other (that is, the woman) sticks her tongue in his mouth. But ultimately the tongue kiss in general is the most pleasurable one, and it holds the key to the mystery of the sweetness experienced when kissing. The reason for this is that, when the lovers suck each other's tongues, "they inadvertently drink each other's spirits."[9] The term "spirit" (*spirito*) here refers to a long tradition that goes back to the Middle Ages and to the work of Averroes. This "spirit" that one drinks is an entity that mediates between the purely material and the purely spiritual. It is a most subtle vapor produced in the bloodstream that travels through the arteries pumped by the heart, reaches the extremities of the body, and comes out through the pores and through the eyes in the form of rays that ignite erotic desire. Some parts of the body that are humid, or spongy, have a much higher concentration of spirits than others, says Patrizi, and that is why in the tongue kiss, as well as during coitus, the exchange of spirits reaches an extreme degree of intensity. Erotic pleasure is a product of the preliminary physical contact when spirits collide. Spirits are thus the ground zero of tangibility since before them there is nothing but the soul, an utterly intangible entity in Patrizi's Platonic worldview. This makes the romantic kiss, the tongue kiss, the final frontier between body and soul. Even though this notion of the kiss as a bridge between soul and body is not original to Patrizi, the novelty of a work like *Delfino* is that, apart from being one of a kind in its speculative treatment of the matter, it approaches the topic from an experiential perspective and bases its claims on notions derived from medicine and metaphysics. This makes Patrizi the first (and only?) fully fledged philematologist, an intellectual hapax.

Leaving aside something as clearly *demodé* as the physiologico-metaphysical doctrine of the "spirits," the unique kind of pleasure felt during the romantic kiss is in itself a matter worthy of consideration. Young Delfino and the seducer Johannes were both on to something when they asked themselves what it was about the kiss that made it so pleasurable, soft, and sweet, and, at the same time, so electrifying and erotic. Whereas Platonists such as Patrizi and, earlier, Castiglione believed the reason to be that the kiss is a meeting of souls, for many lyric poets who engaged with the topic there were other additional reasons. The kiss is, in fact, a hazy occurrence during which body and soul come into contact with each other, but for that very reason it is also at the border between life and death, between being and nonbeing. During the erotic kiss lovers breathe into each other's mouth and inhale each other's breath. The metaphor of the kiss as a bridge that connects life and death is one of the central motifs of a particular lyrical tradition that put the kiss at the center of erotic pleasure. This tradition is born in the Hellenistic period with the works of Callimachus, Theocritus, and other poets later compiled in the *Greek Anthology*; it continues in Rome with Catullus, Ovid,

Propertius, and Tibullus, and has a rebirth in early modern Europe. In the sixteenth century the trope acquired such popularity that a Dutch poet named Jan Everaerts, but better known as Johannes Secundus, decided that it merited a subgenre of its own. That is how he invented the *basium*, a short lyrical piece that deals exclusively with kissing. The *basium* differs from other lyric subgenres like the ode, the elegy, the madrigal, the *strambotto*, and the *amoretto*, not by its form (it doesn't have a fixed metrical structure, actually), but by its content. In fact, the *basium* is not simply a poem about kissing; it is also conceived by Secundus as a kiss from the poet to the reader. For Secundus, to sing about kisses is to give kisses. This might make the *basium* the only form of art conceived as essentially tactile.

Secundus's main work, the *Basia* (*Kisses*), is a collection of nineteen *basia* and it was first published posthumously in 1539. In it, the poet conjures the astounding variety of kisses in a polished and vivacious Latin. When Venus invents the kiss—the myth is told in the first poem—the word Secundus uses is *basium*, a kiss given with open mouth. *Basium* is the most common term for the erotic kiss and it differs from *osculum*, the standard Latin word for "kiss." *Basia* can be dry or humid, long or short, cool or warm, with or without tongue, and they may, or may not, involve sucking. But Secundus also refers to the *suaviolum*, a kiss that is short, soft, and subtle. And the *basiolum*, also a short kiss, one that includes a touch with the tip of the tongue. And then there's the *morsus*, or "biting kiss." All these kisses can be given at different speeds, and with varying degrees of intensity and frequency. If we apply this wide range of philematic options to the complex geography of the human body with its folds, its twists, and its creases, with its flatlands, valleys, and hills, with its deserts and its jungles, with its caves and its promontories, then the possibilities become endless. And it is not just the vocabulary. The metric variety in the *Basia* reflects this astonishing versatility: elegiac distich,

Phalecian hendecasyllable, dactylic hexameter, iambic dimeter, trimeter, tetrameter, and more. Like Patrizi a few decades later, Secundus also considers the tongue kiss to be the crowning achievement of osculation. Not only because it is the one that is given and received simultaneously, but because, in it, the tongue works in tandem with the lips, and mouths alternate between kissing, sucking, licking, and biting. In the dark and humid confinement formed by the mouths, the tongues coil around one another and swirl like battling snakes, quickly opening a space of intense pleasure which, for the poet, reveals itself as a liminal zone between life and death, time and eternity, body and soul. In *Basium* 5, Johannes Secundus describes this with titillating eloquence:

> And your trembling tongue vibrates here and there,
> And my plaintive tongue you suck on here and there,
> Exhaling into me the breath of your soft soul,
> Your delightful, soft, melodious, moist soul
> Neaera, nourisher of my miserable life,
> Drinking my frail, burning soul,
> cooked in the intemperate vapor,
> cooked in the heat of an impotent chest . . .[10]

First it is her trembling tongue that vibrates here and there, then it is his plaintive tongue that she sucks on here and there (*Et linguam tremulan hinc et inde vibras / Et linguam querulam hinc et inde sugis*).[11] The five instances of ploce attach these two lines to one another in a serpentine embrace that reproduces the whirling and twirling of the tongues inside the mouths. With the shared moisture comes the exchange of souls articulated by two gerunds that produce the effect of movement back and forth typical of the romantic kiss. She is exhaling (*asparans*) into him the breath of her delightful, soft, melodious, moist soul. She is drinking (*hauriens*) his "frail, burning soul, cooked in the intemperate vapor, cooked in the heat of an impotent chest." With a succession of adjectives the poet describes her soul with remarkable attention to its distinct sensorial qualities, and he does so by summoning not only tactility and taste (*suavis, mollis, humida*), but also hearing: *dulcisona*. Sight seems to be utterly absent from this intimate negotiation.

In *Basium* 4 he says: "let me devour your mouth a thousand times and I will become immortal;" and in *Basium* 10, as he describes the tongue kiss: "to suck your trembling tongue with my plaintive lips, and mix two souls in one mouth, and then when our love languishes to the point of resembling death, to diffuse our pilgrim bodies one into one another."[12] The kiss appears thus as a *petite mort* that is followed by immediate resuscitation. If, as Plato claimed, philosophy, understood as an attempt to separate body and soul, is but a practice in dying, in the fantasy of these poets kissing is a strictly philosophical activity.[13] For Plato and Socrates, the only suitable medium for philosophy is the dialogue. And isn't the kiss an alternative form of dialogue that dispenses with words, although not with the tongue? Secundus would have most likely agreed with Patrizi and with Kierkegaard when they chided philosophers for neglecting the philosophical potential of the kiss.

Soon after Johannes Secundus's premature death at twenty-four, the *basium* is translated into the vernacular and becomes a staple of early modern lyric poetry. Some of the greatest poets of the sixteenth and seventeenth centuries adapted Secundus's images and motifs into their own work and even wrote their own *basia*. Ronsard, Du Bellay, and Belleau in France, Marino in Italy, Sidney and Shakespeare in England are some of the most prominent examples, but there are far more. The fashion of the kiss poem was pervasive and it extended all the way to the Romantic period. In his youth, Goethe was so smitten by Secundus that his friend Herder gave him the nickname Johannes Tertius. In England, Byron and Shelley too engaged with various kiss motifs, and later in the century so did Swinburne and Tennyson, among other Victorian poets. But going back to the Renaissance, among Secundus's imitators and admirers no one captured the intense psychological complexity of the kiss like Maffio Venier. Venier, who lived in the second half of the sixteenth century, was a young Venetian rakehell and a good friend of Francesco Patrizi. This poem written in Venetian opens up new ways to think philematologically:

> Kiss me, my darling, and make me die
> With a strike of that little tongue of yours,
> Kiss me, darling, kiss me, murderess,
> This kissing is so lovely, do kiss again.
> Now kiss me one more time, kiss all the time,
> Kiss always, always kiss, kiss to perfection,
> Kiss me, bitch, a thousand times an hour.
> Kiss me, my soul, kiss me so much
> Make me miserable with kisses,
> Kiss, and after you've kissed me well, kiss me some more.
> Kiss me as much as it's humanly possible,
> Kiss, kiss me until I speak no more:
> Until I lie dead in the grave.[14]

The poem first strikes us for its violence and for its repetitiveness. Interestingly, there is neither crescendo nor climax. It begins and it ends in the same exasperated and imperative tone; the circular structure suggests that the kiss has not begun and it will not end. The poet plays a stern, active role, commanding and directing, and a passive role: he demands to be kissed. This bespeaks the reversal and interchangeability of roles during kissing, and is also reminiscent of Delfino's conclusion that the kiss received is the sweetest one. The images of death and dying are both veiled references to the orgasm as well as examples of the classical motif of the exchange of souls that happens during the kiss. As for the epithets addressed to the beloved ("darling," "murderess," "bitch," "my soul"), they illustrate the feverish passion that oscillates between abuse and affection and that drives the poet to a state of erotic schizophrenia. In Venier's poem, the interstitial space of the kiss with its aesthetic complexity, its ambiguities, and its preliminary nature agitates the lover to no end. Finally, the fast and iterative prosody imitates the pacing and rhythm of the kiss, but also that of sexual intercourse. Unlike sex, however, that inevitably ends with the orgasm, kissing is an activity that knows no climax and that could very well be practiced *ad infinitum* (though not *ad nauseam*).

From Catullus to Shakespeare, the motif of the kiss as a potentially never-ending negotiation, a ludic bargain, has been a constant in amatory poetry. In Secundus's *Basium* 6, the poet pesters his beloved every time she wants to leave, asking her for a few more kisses that he will return, with interest, at a later time. Shakespeare too engages with this tradition, especially in his narrative poem *Venus and Adonis* and, most famously, in plays like *Romeo and Juliet* and *Antony and Cleopatra*.[15] The potential inexhaustibility and the chaste nature of the kiss make it easier to indulge in, less risky, and thus the perfect form of erotic training for adolescents. But the kiss is propaedeutic not only as a preamble to sex (Philip Sidney calls it "breakfast of love");[16] in other words, the kiss is not invariably subject to the coitus. The kiss is *essentially* propaedeutic because it constitutes in itself an erotic and sentimental education, a continuous and meticulous sensorial exploration of another body during which the very "chemistry" between lovers is tested.

If Patrizi thinks of the kiss as the space where body and soul conflate and Johannes Secundus imagines it as a hazy frontier between life and death, in Venier's frenzied verses and in the motif of the kiss as exchangeable commodity, the kiss, given its potential to be prolonged indefinitely, appears as a threshold between time and eternity. As noted before, unlike intercourse, the kiss is aclimactic, and for lovers, whose biggest source of anxiety is separation and whose most dreaded fear comes from the prospect of an ending, the microcosm formed within the cavernous confines of two mouths closed onto one another, amidst trembling lips and tongues that roll and scuffle; for lovers, I say, this purely haptic world moisturized by the shared flow of saliva is a *locus amoenus*, a bubble outside of time where the fiction of a love that lasts forever can endure safely.

But the kiss is also didactic because it allows for interchangeable roles. Both parties are equally equipped to explore the nuances of desire and to be puzzled by the conundrum of satiety. In *Astrophil and Stella*, Sidney says that kiss teaches "the mean at once to take and give."[17] Precisely, in the kiss both parties are at once active and passive because the lovers penetrate and are penetrated, alternatively or simultaneously, by means of the tongue, a phallic substitute that, unlike the male organ, works both erect and flaccid. And because it doesn't require fixed, pre-established roles, the kiss is a nonhierarchical, horizontal, and perfectly reciprocal endeavor. No party can claim complete domain over the other: whoever bites will be bitten, whoever licks will be licked, whoever penetrates will be penetrated. This makes it the fairest and most egalitarian (the most democratic!) of all erotic practices. I said before that the tongue kiss is a nonverbal form of dialogue (it is indeed telling that the only philosophical work ever to discuss it is a dialogue); this too is indicative of its horizontality.

And yet the erotic kiss, even conceived as a liminal space, as an instance of transition, or mediation, as an initiation, or as an education, continues to resist the claws of language and escape both thorough description and conceptualization. Naturally, it is a slippery thing. Not necessary for reproduction, culturally inflected and, as it has been shown, by no means universal, the kiss plays an exceptional role in human relationships, not least because of its daunting versatility. Even though sexual intercourse also allows for a wide variety of possibilities, the orgasm limits it irredeemably. Free of all climactic constrictions, the kiss is perhaps the most pure manifestation of erotic desire because when it seems that it is nearing satiety, it changes pace, switches gears, or becomes something else that is redirected and continues to yearn and burn like a perennial flame. But the erotic kiss reflects the nature of desire in a more specific sense. Tongues intertwine with tongues over and again, clockwise and counter-clockwise, they advance and they recede, they dilate and they shrink, they lick each other, they chase after one another, they clash and they wrestle, they lick teeth and lips, and lips in turn open and close, suck and absorb while being sucked, while being bitten, and teeth do bite, they crash against teeth (an unpleasant turn that soon shifts and is forgotten), and so on and so forth. The constant and gymnastic mutability of the kiss mimics the dynamics of erotic desire that, like an invisible Proteus, is elusive, chameleonic, and hyperactive. Both the kiss and desire itself are instances of mediation that exist—or, rather, that *happen*—in liminal zones: the kiss, in the no man's land between body and soul, time and eternity, life and death, love and lust, masculine and feminine; desire, in the nondescript frontier between its fathomless origins and its unattainable object.

A scene from Andrei Tarkovsky's first feature film, *Ivan's Childhood* (USSR, 1962; see fig. 3), portrays the liminal nature of the kiss in a way that is at once inspiring and disenchanting. During World War II, somewhere along the Eastern front, a captain and an army nurse go for a walk in the woods. As they get to a ravine, the man goes to cross it first and with a foot on each side offers to carry the woman across. He then lifts her up and, as her legs hang in the air, he kisses her. The camera captures the moment from a low angle, deep inside the ravine, his legs wide open, her legs hanging straight in between them, like the clapper of a swinging bell, a suggestive evocation of the gender reversal enacted in the kiss. The lightness of her legs contrasts with the firmness of her upper body that, held

3.
Ivan's Childhood
(dir. Andrei Tarkovsky, USSR, 1962).

in the man's firm embrace, seems to be petrified in the kiss. It is a stolen kiss, a seemingly insignificant act of violence within a much more dramatic context, that she is too unaffected to fight. This scene functions as a digression, an attempt to take a break from the anguish and brutality of the war. But the mood created by the war impregnates everything and the attempt is ultimately unsuccessful; the characters are tense (she's too unnerved, he's too aggressive), the ravine reminds us of a trench (or of a mass grave), and at the moment of the kiss we hear shots fired in the distance. The kiss here is a dystopian *locus amoenus* that bridges a divide between war and war, and Tarkovsky's stages it on a triple frontier: the no man's land of the Eastern front, the deep ravine, and the two elevated bodies.

The kiss as a tool of erotic domination and manipulation is the theme of Guy de Maupassant's 1882 epistolary short story "Le baiser" ("The Kiss"). In it, a young woman who is having marital trouble receives a letter from her aunt Colette with advice on how to reclaim and retain the attention of her sexually distracted husband. The key, says Aunt Colette, is to master the sense of touch and, in particular, the kiss. Whereas men dominate by means of brute force, women do it through caresses that soften and tame, the aunt begins. Softness always prevails over hardness. Softness is adaptable and enduring; rigidity withers and breaks. The aunt reminds her niece that Venus always vanquishes Mars. "Love, my dear, is made up of imperceptible sensations; we know that it is as strong as death, but also as frail as glass,"[18] she writes. Attention to detail is the key to success in any art or trade; so too in the art of love. Those "imperceptible sensations" are crucial: the lover must develop a specific form of hypersensitivity and learn to appreciate what he or she has hitherto taken for granted and overlooked (or, rather, "overfelt"). In order to achieve this, a balance must be reached between reason and passion that permits one to get carried away by erotic pleasure without losing control of the situation, and without neglecting the complexities of the amorous encounter. The young couple must attend to the variety of tactile shades and tinges—the *nuances*. They must dedicate time and patience to the erotic ritual, because the biggest enemies of eroticism are haste and abruptness. And, in order to educate someone in such nuances, Aunt Colette concludes, there is nothing better than a good kiss:

> Do you know whence comes our real power? From the kiss, the kiss alone! When we know how to offer and give up our lips we can become queens. The kiss is only a preface, however, but a charming preface. More charming than the realization itself. A preface which can always be read over again, whereas one cannot always read over the book. Yes, the meeting of lips is the most perfect, the most divine sensation given to human beings, the supreme limit of happiness. It is in the kiss alone that one sometimes seems to feel this union of souls after which we strive, the intermingling of swooning hearts, as it were.[19]

The call to offer our lips resonates with Sidney's description of the kiss as a teacher of giving and taking. In the kiss we learn to interpret amatory roles, and we discover the erotic idiosyncrasies of the other, but also of ourselves. According to Aunt Colette's interestingly subversive heteronormative position, the ultimate goal of the kiss is for the woman to control the man by pleasing him. The notion that the kiss happens in a liminal space is suggested with the evocative "preface." This in-betweenness of the kiss gives it the elasticity that allows for its didactic potential; one can go back to it time and again, whereas "the realization itself" is exhausted with the climax (not to mention it is logistically more complex and subject to legislation). Finally, the trope of the kiss as the union of two souls also appears in the aunt's account, albeit filtered through the sieve of secularization. There is no sense of an actual merger here, but mere hope and yearning for a form of contact that transcends the body. Not unlike the tickling that amputees sometimes feel in their phantom limbs, the pleasure of the kiss comes from that metaphysical tingling of the intangible against the tangible. This strange sensation that the aunt characterizes in physiological terms as an "intermingling of swooning hearts" refers to the liminal status of the kiss that Patrizi explained through the extravagant doctrine of the "spirits," and the sensation that the poets elicited with their mystical and erotic metaphors. In the kiss, his heart—hardened, cold, jaded—will become hers—soft, tender, warm. She will reconquer his attention and Venus will once again triumph over Mars. The aunt concludes: "Therefore, my dear, the kiss is our strongest weapon, but we must take care not to dull it. Do not forget that its value is only relative, purely conventional. It continually changes according to circumstances, the state of expectancy and the ecstasy of the mind."[20]

Aunt Colette implies that to be a good kisser means to be versatile, at once dominating and malleable, attentive to the nuances, particularities, and cravings of the other body, capable of improvising and changing pace. The only bad kisser is the mechanical kisser, the one-trick pony of osculation who kisses everyone always in the same exact manner, an example of which I found in the third volume of *My Struggle*, where Karl Ove Knausgaard describes a long and painful kiss between two pubescent teens. The narrator takes his brand-new girlfriend, Kasja, for a walk in the woods where he plans to kiss her for the first time. As they get to a clearing, he pulls out a pocket watch and proposes that they kiss until they beat the record of ten minutes held by his friend Tor. This is what follows:

> I put the watch down on the ground, it was eighteen minutes to eight, I noted, placed my hands on her shoulders and gently leaned her back while pressing my lips against hers. When we were both lying down I inserted my tongue in her mouth, it met hers, pointed and soft like a little animal, and I began to move my tongue round and round inside. I had my hands alongside my body, I wasn't touching her with anything except my lips and my tongue. . . . I concentrated on getting my tongue to go round as smoothly as possible while the thought of her breasts, which were so close to me, and her thighs, which were so close to me, and what was between her thighs, under her trousers, under her knickers, was seared into my consciousness. But I didn't dare touch her. She lay with her eyes closed rotating her tongue around mine, I had my eyes open, groped for the watch, found it and held it within reach. Three minutes so far. Some saliva ran down from the corner of her mouth. She wriggled. I pressed my groin against the ground letting my tongue go round and round, round and round. This wasn't as good as I had imagined, in fact it was quite strenuous. . . . Our mouths were full of thick saliva. . . . Mmm, she said, but this was not a sound of pleasure, there was something wrong, she stirred, but I didn't let go, she moved her head while I continued to rotate my tongue. . . . Nine minutes. The root of my tongue ached. More saliva from the corners of our mouths. My dental brace

> occasionally knocked against her teeth. Actually we didn't need to continue for more than ten minutes and one second to beat Tor's record. And that was now.... But we could beat him by a large margin. Fifteen minutes, that ought to be possible.... But my tongue ached, it seemed to be swelling, and the saliva, which you didn't notice much when it was hot, left you with a slight feeling of revulsion when it ran down your chin, not quite so hot.... At exactly three minutes to eight I took my head away. She got up and wiped her mouth with her hand without looking at me.[21]

The reader is not surprised to find out that, the next day, Kasja breaks up with young Karl Ove. The mechanical kiss cancels all possibility of eroticism: it is the anti-kiss. Someone has told the young narrator that kissing consists in inserting the tongue and rotating it incessantly with both authority and perseverance. His reticence to touch her isolates the kiss and reduces it to lips and tongue exacerbating its cold, mechanic feel; the young Karl Ove is still not aware that kissing is a pastime in which the whole body takes part. He is unsettled by the proximity of her body, by mental images of her breasts and her thighs, of her pants and underwear; shocks of libidinal anxiety run through him like electricity underwater, but he represses his tactile impulses subjugating the proprioceptive instincts of his erotically charged hands, legs, and groin, and instead concentrates all of his efforts in the pointless and monotonous circumvolution of the tongue. Above all, the record-breaking kiss is unable to give either of the two youngsters any pleasure because it is the mindless repetition of a protocol. Kasja and Karl Ove are following instructions; they have goals and methodologies that have been dictated by others and this prevents the wonder of genuine philematic interaction from happening. Their kiss is a monotone of simultaneous monologues, rather than a nonverbal dialogue. What's missing more than anything is improvisational skill. In his tenth *basium*, Johannes Secundus compares kissing with a musical duet. As both parties improvise and show off their dexterity and virtuosity, they must be acutely aware of one another and interact with ingenuity. As in improvisational couples dances such as the tango—another form of nonverbal dialogue in which a light squeeze or a change in the grip is crucial to the flow of the pace—the only golden rule for kissers is to develop a hypersensitivity that allows them to effectively engage with the other, acting and reacting in accordance with the subtlest haptic feedback. Philematology must then be a descriptive rather than prescriptive form of knowledge; more similar to biology and chemistry than to ethics and philosophy. And given that its body of work is made out of a comprehensive catalog of kisses drawn from experience, it seems closer to the natural sciences than to the humanities. Still, the most suggestive depictions of the kiss come to us from literature. In the seventh chapter of *Hopscotch* (1963), Julio Cortázar recounts a torrid kiss and, in doing so, summarizes with exuberant prose two thousand years of osculatory poetry and philematological wisdom:

> You look at me, from close up you look at me, closer and closer and then we play Cyclops, we look closer and closer at one another and our eyes get larger, they come closer, they merge into one and the two cyclopses look at each other, blending as they breathe, our mouths touch and struggle in gentle warmth, biting each other with their lips, barely holding their tongues on their teeth, playing in corners where a heavy air comes and goes with an old perfume and a silence. Then my hands go to sink into your hair, to cherish slowly the depth of your hair while we kiss as if our mouths were filled with flowers or with fish, with lively movements and dark fragrance. And if we bite each other the pain is sweet, and if we smother each other in a brief and terrible sucking

> in together of our breaths, that momentary death is beautiful. And there is but one saliva and one flavor of ripe fruit, and I feel you tremble against me like a moon on the water.[22]

As the distance between the lovers is reduced, four eyes merge into two cyclopean glances that then become one; sight progressively gives way to touch, hearing (*un silencio*), smell, and taste. The conflicted relationship between the senses during the kiss—at times battling, at times collaborating—is a motif in the *basia* tradition. Cortázar also echoes others: hands are summoned, the pain from biting is sweet, there is a dark fragrance, and the momentary death during exhalation is bliss. Finally, as they drink each other's saliva, the lovers become one; but then one of them trembles and reverberates against the other. They are two again and, even though the kiss continues, Cortázar ends with a specular image (the slightly rippling reflection of the moon on the water) that drags us back to the precarious realm of the visual.

Two final questions before I shut my mouth. The first one relates to heteronormativity and the erotic kiss, the second with the connection between kissing and eating. For Patrizi, Kierkegaard, and Maupassant, the kiss was conceived as a strictly heterosexual matter. Yet in the poets of the *basium*, roles and genders were often reversed and became blurry in the liminal zone of the kiss. And while Patrizi and Kierkegaard claim that the only kiss that can be of any philosophical interest is that between a man and a woman, one of the oldest kiss poems that we know of was not only attributed to a philosopher—Plato—but it evoked a kiss between two men. With this antecedent in mind, perhaps it need not surprise us that one of the most lucid accounts of the nature of the kiss in the early modern period comes from the pen of an infamous apologist of homosexuality. Antonio Rocco, who understood the romantic kiss as not only pleasurable but also didactic (*dulce et utile*), discusses the issue establishing a strong connection between the tongue kiss and rhetorical skill. He was not, however, the first to tackle the topic of the male-to-male kiss. In *L'impresa* ("The imprint"), a rare and still unpublished 1569 *dialogo d'amore*,[23] Cesare Trevisani briefly broaches the issue. Trevisani distinguishes between four types of kisses: the sacred kiss (given at mass and also to sacred objects), the civil kiss (given to kings, royalty, and nobility), the loving kiss (the kiss between friends, family members, or spouses), and the lascivious kiss (the kiss between lovers). Trevisani is not alone in distinguishing the kiss between spouses from the kiss between lovers; in a seventeenth-century distich, the English poet Robert Herrick does it with succinct elegance and humor: "Kissing and bussing differ both in this: / we buss our wantons but our wives we kiss."[24] So too Johannes, Kierkegaard's proto-existentialist Casanova, claimed that the spousal kiss is the anti-kiss, devoid of all eroticism, mechanic, and utilitarian. But Trevisani's novelty is that he admits that the lascivious kiss can be both homosexual and heterosexual.[25] Hardly a century after Trevisani wrote this, Antonio Rocco published one of the most vilified and demonized works of early modernity, *Alcibiades the Schoolboy* (1652). This dialogue, which prefigures the main ideas of some of the most influential eighteenth-century French libertines, deals with a pederast (in the role of the teacher) who puts all of his rhetorical skills to the service of seducing a young boy (the pupil) by persuading him that homosexual sex is not only more pleasurable but also more divine than heterosexual sex. Rocco argues that sexual pleasure is a natural phenomenon and that depriving oneself of it (depriving oneself of any pleasure, for that matter) constitutes an affront to nature itself. In short, sodomy is not *contra natura*, what is against nature is not practicing it. The kiss is among the issues broached by the teacher. He explains that the tongue's home is the beloved's mouth and her reason for being, her *telos*, is the kiss. And in a cunning apology of the dialogue as the most effective philosophical genre, the teacher adds: "The eloquence you shall learn from my doctrine . . . will never be absorbed by you if our tongues are not intertwined. The hand educates the hand, the mind educates the mind, and the tongue educates the tongue."[26] After saying this, he kisses the pupil, inserting his tongue in the young man's coy mouth. This is when true learning begins; we, the readers, are left out. Paraphrasing the pseudo-platonic distich, Rocco concludes that the pupil's soul concentrated all on his lips and "his whole life was nothing but a kiss."[27]

From the *teleology* of the kiss to its *etiology*. I suggested that the erotic kiss is a nonverbal form of dialogue, a haptic conversation. Sexual intercourse is too, and when the two happen simultaneously a paroxysm of erotic reciprocity ensues that is perhaps the *non plus ultra* of physical intimacy. But where does the impulse to kiss come from? According to Wilhelm Wundt, the erotic kiss between man and woman could be traced back to the primitive belief that the exchange of breath was a necessary condition for procreation;[28] as if the two exhalations combined produced the magic sparkle that ignites life. This belief goes hand in hand with the notion that procreation is the literal expression of the metaphor of two bodies becoming one during intercourse. The kiss too has traditionally symbolized for many cultures the fusion of two into one. But there is another theory that connects the origins of the erotic kiss with the act of eating, and with a form of oral fixation by which we draw toward our mouths anything that is in any way appetizing, or appealing.[29] In these lines from book 4 of *On the Nature of Things*, Lucretius seems to hint at both explanations for why people kiss, combining them into one and harnessing them to his overall pessimistic outlook on sexual desire:

> Boften [lovers] set their teeth in the lips and crush mouth on mouth. . . . Lastly, when clasped body to body they enjoy the flower of their age, at the moment when the body foretastes its joy and Venus is on the point of sowing the woman's

> field, they cling greedily close together and join their watering mouths and draw deep breaths pressing teeth on lips . . . [but] they [cannot] penetrate and be absorbed body in body.[30]

For Lucretius, as the previous essay showed, romantic love is an exasperating chimera. The desire to merge with the other and the utter impermeability and immutability of the atoms that conform us result in the frantic cycle of ardor and dissatisfaction that makes sexual desire such a disruptive distraction. Given the impenetrable nature of our atoms, even in the closest proximity lovers feel an insurmountable distance that separates them; and that is what makes them try to devour one another. This doesn't differ substantially from what the *basia* poets, Patrizi, and other Renaissance love philosophers believed. For them too the kiss was a manifestation of intense spiritual hunger and of deep physical yearning. This yearning centers on oneself, on abandoning one's own body and becoming undone in someone else. At the same time, the focus is the other and the yearning is to possess, to absorb, and to devour the beloved's body. The intrinsic reciprocity of the erotic kiss allows for such synchronicities. But if lovers simultaneously transcend their own bodies *and* devour one another, it is because in kissing they themselves become the kiss. This might be the biggest mystery of philematology. Secundus hints at it in *Basium* 2 when he wishes both lovers would dissolve in each other's kiss, and Brâncuși's *Le baiser* (1913) is perhaps the artifact that most powerfully evokes it. The sculpture shows the lovers of unidentifiable genders as one block of plaster, wrapped in a serpentine embrace. Our attention is immediately directed to the small indentation of their shared mouths. And then it hits us; they are not two lovers, they are one kiss.

THE FRENCH CONNECTION

Between 1767 and 1794, typographer, pamphleteer, pornographer, novelist, playwright, proto-*flâneur*, insomniac, and eponymous hero of retifism Nicolas-Edme Rétif, better known as Rétif de la Bretonne, spent most of his nights walking the streets of Paris. The product of his relentless wandering was *The Nights of Paris*, a collection in eight volumes, published between 1788 and 1794, that defies all classification. Among the myriad of atrocious and hilarious characters that fill the pages of this mammoth work, I would like to highlight one; a man who, during the years of the Jacobin terror would go to public executions, position himself strategically in the middle of the crowd and, at the precise moment when the guillotine fell, grope a woman's rear. This joker (or degenerate, as the reader wishes) was known as *le toucheur*, "the groper."[1] I invoke the figure of the groper as the emblematic symbol of a cultural sensibility, and in the following pages I propose to consider the idiosyncratic fixation that certain French intellectual and artistic traditions have had with the body, the flesh, and the sense of touch.

French artists and intellectuals have focused on the body, on sensibility in general, and on the sense of touch in particular more than any other in the Western tradition. But while poets, painters, and sculptors can represent and evoke the tactile simply by means of their craft, there seems to be an extra degree of separation between the philosopher and all issues pertaining to the senses, namely speculation itself. Melville's prosody, the Homeric dactyl, Lucretius's alliterations, and Johannes Secundus's kisses are only a few examples of how language can reproduce haptic operations. The visual arts also have techniques to appeal to the touch through the eye and to mimic the haptic (foreshortening, linear perspective, close-ups, etc.). Philosophy and theory, however, when approaching such topics are limited by some of their most fundamental prerogatives. Analysis, interpretation, and classification require both a clear distinction between subject and object and a fair distance between them. One could say that to philosophize is in itself an act of separation, of estrangement, of detachment, or removal; and in this sense, it is an activity that resonates with sight rather than with touch. Speculation (from the Latin *speculum*, "mirror") is indeed a game of mirrors. Both Platonism and Aristotelianism betray this visual bias in their conceptual fixations with mediation and representation, with images and models, with taxonomies and categories. It is telling that, in spite of their crucial differences, Plato and Aristotle agreed on the fact that sight was the noblest, most trustworthy of all senses. Touch, however, and the faculties encompassed under the notion of the haptic operate in close proximity, in a space that is opaque and confusing, a space that Michel Henry refers to as "invisible." To theorize about touch would thus seem impossible.

Is it possible to think without distance? Can one speculate, distinguish, classify, hierarchize in the dark? Since early modernity on, a great number of French thinkers have pondered these questions and attempted precisely this. The topic is vast and I must leave out an enormous amount of relevant material, not only philosophical but also from literature and the visual arts.[2] The sense of touch is discussed and evoked in provocative ways in the works of, among many others, Condillac, Maine de Biran, Sade, Balzac, Huysmans, Verlaine, Cézanne, Apollinaire, Rodin, Levinas, Bresson, Irigaray, Nancy, Derrida,

Gentaz, Serres, Anzieu, and Didi-Huberman. I will focus mainly on two approaches to the topic: the first, which initially appears in early modernity and continues through the eighteenth century, can be labeled neo-Epicurean; the second is twentieth-century phenomenology, in particular the work of Michel Henry.

The immanent materialism of the atomistic tradition runs through Western intellectual history since pre-Socratic times until our own days like a deep wound that refuses to heal. Its sensualist epistemology and its emphasis on the various modalities of touch make it one of the richest traditions to help us think about the tactile and the haptic. As for phenomenology, some of the issues addressed in Husserl's late writings, and pursued subsequently by Merleau-Ponty and Michel Henry, touch on the problem of the body as living organism and engage with topics concerning the tactile and the tangible.[3] Although he does not use the term haptic, Michel Henry provides conceptual tools to approach the issue because his project identifies the very origin of experience with a revelation that he calls "self-affection," a form of affect that precedes all sensitivity, and by which the organism recognizes itself as a living being. Henry thus localizes the ground zero of feeling and sensibility in an experience that can be deemed proto-haptic, an ontological proprioception of sorts.

As he develops the notion of self-affection, Henry positions himself on the margins of traditional, Husserlian phenomenology, but he also takes a harsh stance against older schools of philosophy, and especially against atomism. Henry believes that atomism, like most philosophical traditions born in classical Greece, is guilty of epistemic naiveté because it failed to recognize that all inquiry about existence must begin from a self-revelation of life as flesh. This betrays a hasty (or an insidious) reading of Epicureanism, fueled perhaps by the fact that Henry ultimately finds in Christian theology and in the dogma of incarnation the seed for his phenomenology of the flesh. There is a point of contact, however, between Henry and atomists like Lucretius that sets them apart from the traditions started by Plato and Aristotle. Both of them identify the origins of being and the origins of sensation and knowledge with the realm of the invisible. And sight is for them a delusory sense: a sense that is as limited as it is secondary in importance. Notions such as those of idea, form, phenomenon, representation are distractions that keep us from the true, deep roots of life and of experience. For Henry in particular, philosophers must work in the dark, like the blind. In the pages that follow I will adopt this approach. In order to sketch out a cultural sensibility that moves transhistorically, in order to think about the invisible and about the flesh, I must proceed blindly, groping about in the dark, touching on a passage from troubadour poetry here and a poetic kiss there, brushing against an early modern science fiction novel here and a philosophical letter there. And since French culture has been fertile ground for gropers from its earliest days, it is back to the most distant past of the vernacular tongue that I now jump to begin my fumbling journey—all the way back to the medieval dawn of French literature and to an iconic scene, a scene of harrowing darkness in which the human body is violently ripped open to inaugurate for posterity a millennium of explorations through the winding and intricate world of touch.

Composed in the late twelfth century, Chrétien de Troyes's *Lancelot, the Knight of the Cart* tells the story of the ill-fated affair between King Arthur's wife Guinevere and his most trusted companion. The villain Maleagant has kidnapped the queen, and Lancelot, eager to prove his devotion to the king, sets out to rescue her. After a long journey beset with difficulties and obstacles that include dead horses, spiteful dwarves, a bridge of swords, a treacherous courtesan, and one too many bloodthirsty warriors, Lancelot vanquishes Maleagant and rescues the queen. The moment the ordeal is over new trials begin, as the knight and the queen, bursting with joy and excitement, fall madly love with each other and agree to meet that very night. Fearing the inability to restrain her desire, Guinevere asks Lancelot to go outside her window, where thick iron bars will separate their bodies and yet allow them to talk, hold hands, perhaps exchange a kiss, or two, but no more. "You cannot come inside or be with me: I shall be inside and you without,"[4] says Guinevere. Night falls and, under a moonless and starless sky the lamps are put out and the world is wrapped in a blanket of absolute darkness. Blind, guided solely by his desire, Lancelot arrives at the queen's window. They hold hands and speak words of love. It's not long before the inflamed lovers decide that they need to be together. In the pitch-black night, their bodies palpitate, and their skin burns with passion. The queen yearns to be "quite naked next to him, in order to enjoy him fully" and Lancelot cannot but oblige her.[5] No matter how strong the bars may be, Lancelot says, they will not prove an obstacle for his desire; and he proceeds to bend them with superhuman strength, slicing off his fingertips with the cold, sharp iron. But his body, electrified by desire, does not feel any pain; he doesn't even realize that he is wounded. Dripping blood, he reaches the bed where Guinevere waits for him in exultation. The lovers spend the night intertwined and the next morning, Lancelot leaves after bending the bars back into place. It is only when he arrives in his quarters that he notices his mutilated fingers; and this happens at the exact same moment that a page wakes up Guinevere screaming upon seeing her sheets drenched in blood.

Several aspects of this episode anticipate the haptic sensibility that French culture would develop in an ever more self-aware way during the course of the following centuries. The sexual symbolism of the bending of the window bars is quite obvious but not for this less interesting. Guinevere had warned Lancelot that he would not be allowed to enter. When he comes to her window, though, the poet stages the scene adapting an old device of lyric and troubadour poetry, the *paraclausithyron*. In the classic *paraclausithyron*—Pyramus and Thisbe, for example—lovers are inexorably separated by a wall, or a door, and they can only hear, or perhaps catch a glimpse of, each other, but they are denied touch. In this case, however, the bars allow the lovers to hold hands. This initial contact unleashes a tactile frenzy, possessed by which Lancelot is able to free his superpowers, and inflamed by which Guinevere forgets her marital duties. When he forces the iron bars, Lancelot mutilates his fingertips but doesn't feel anything; the surface of his body has been numbed by the erotic affect that moves him toward his final goal, which is, paradoxically, an explosion of physical pleasure. In order to be able to feel intense pleasure, he blocks all feeling of intense pain. Three layers of sensorial and affective experience, which correspond with different instances of the haptic, can be deduced from this scene: first, there is the superficial layer of the antalgic body that can be touched but doesn't feel. Second, this superficial, anesthetized layer imposes its will onto the world and surmounts obstacles (it bends the bars), which means that even without feeling it can perform haptic faculties such as putting pressure and maintaining balance. Finally, there is the deeper layer of the inner body and its pure affect bursting with sexual desire that drives the action.

The bending of the iron bars is also an anticipatory metaphor for the lovers' first sexual encounter. It is Lancelot's body, however, that is torn and opened. He bleeds, not her who has already been deflowered by King Arthur. The blood spilled on the bed, mixed with sweat, tears, and other erotic discharges, is the concrete trace left by the explosive conjunction of the two electrified bodies; and it is also the visible instantiation of the operations performed by the multiple haptic faculties that partake in the erotic phenomenon. The shameful and incriminating stain that soils the bed is, thus, the concentrated sediment of an overwhelmingly complex affective outburst. The next morning, when questioned by the horrified page, a baffled Guinevere explains that she suffered a severe nosebleed during the night.

The first time I read *Lancelot* this passage had an extraordinary impact on me. As it sometimes happens with what we read, in time the plot and the details evaporated almost completely from my memory and all that was left, shining dimly with quiet endurance was just a feeling—what I felt when I came to this scene. And that is because, as I read, the violence of the images and the euphoria of the star-crossed lovers became palpable. Lancelot's maimed fingers, his raw flesh throbbing to the rhythm of sex like an exposed extension of the heart, the blood (how much blood I wondered), the queen's bed exploited until morning, the sheets soiled by the dense concoction of bodily brew, everything materialized for me with such density that it could only be experienced as tangible. Chrétien de Troyes's verses, composed almost a thousand years ago, had worked their magic. Years later, I found in the thought of Michel Henry strong resonances with this medieval scene.

In his late work *Incarnation: A Philosophy of the Flesh* (2000), Henry proposes a radical subversion, or rather a reversal (*renversement*) of phenomenology. He argues that the notion of intentionality, the main conceptual pillar of the tradition born in the works of Brentano and Husserl, far from being the most primordial instance in the relationship between the self and the world, refers to an operation that is already subsidiary. Precisely because it involves a duality—the self, experienced as life, and the world, experienced as representation—intentionality requires distance and is subordinated to the more primal experience of one's own body, or, in Henry's words, one's "flesh" (*chair*). The real cornerstone and ultimate condition of possibility of all experience and of all thought is what Henry calls "self-affection" (*auto-affection*).[6] By this Henry means a prelinguistic, presensorial, and yet unmistakable and overpowering experience of being alive, of being an individual, a concrete, pulsating entity. This experience Henry associates with the ability to move (a proto-kinesthesia, one could say), and links with the most primal expression of animal volition, the mere capability of moving one's body at will that Henry refers to as the "I can" (*je peux*). This liberty of movement is confronted with a "resisting continuum," a first line of external opposition imposed on the body that allows for a double revelation. In Henry's words:

> When the resisting continuum puts up an absolute resistance to the original powers of our original corporeity, this continuum defines the reality of the bodies that make up the "real" universe. When on the contrary it gives in to these powers, the reality of our organic body is revealed in it.[7]

This original collision between the body and everything that transcends it reveals both the reality of the body as organism (when we overcome the resisting force) and the reality of the outside world (when the resisting force prevails over us). I now proceed groping around in the dark. The scene of the iron bars bending in *Lancelot* captures the meaning of this double mechanism that, according to Henry, precedes all sensation and all knowledge. Lancelot's unbridled *eros* functions as self-affection, origin of the "I can" and primal drive that initiates all action and movement. The bars, initially characterized by Guinevere as unsurpassable, represent first the resistance that reveals the reality of the outside world, and then, as Lancelot violates them, the resistance that reveals the body as organism, of which the hero promptly makes use in the queen's bed.

And then there's the darkness. As Lancelot makes his way to Guinevere's quarters he does so under a night sky that hangs low with heavy coal-black clouds. The air is so dense one can practically touch it and Lancelot moves guided by the internal compass of his erotic drive. Guinevere's touch through the window inflames him and he bends the bars. He hurts his fingers, yet doesn't notice it. He bleeds over the bed and neither he nor the queen see the blood. Vision is altogether absent from this first meeting of the lovers. Instead, everything is haptic: things throb, there is movement and contact, there is pressure, laceration, collision, impact, rubbing, penetration, secretion, excretion. With sight canceled completely, Chrétien de Troyes reduces the space and the action to the outbursts of two intertwined bodies, a thriving cumulus of life. For Michel Henry, the reversal of phenomenology and the experience of self-affection are only possible in the sphere of the invisible:

> Is not phenomenology a philosophy, and philosophy a thought, a thought that comes about through vision? . . . If life is invisible, then, how is it possible to have access to it in thought, and how is a philosophy of life still possible? . . . Living, we are beings of the invisible. We are intelligible only in the invisible, and on the basis of it.[8]

If thinking is made possible only by distance and sight, a philosophy of the body as raw, living flesh must be a haptic endeavor. We are, first and foremost, invisible flesh and this is grasped through self-affection. Henry's phenomenology of the flesh rests upon the conviction that one cannot know the phenomenon of life by thinking about it, but only as affect, in the flesh. As if paraphrasing and correcting Marx's incendiary thesis 11 on Feuerbach, Henry seems to be saying that philosophers have hitherto only *seen* the world in various ways when the point is to *live* it.

In the narrative proposed by Husserl in his late writings, which Henry adopts and continues, the representational fixation of Western philosophy (the seed of what others will later call ocularcentrism) is exacerbated in the early modern period with the writings of Galileo and Descartes. According to this rather inexact and somewhat simplistic, albeit thought-provoking, narrative, modern science covers the world as we experience it and live it, with a veil of ideas thus turning it into something essentially quantifiable and commensurable. The reality of the world is then accessible to us by means of observations, speculations, inductions, and deductions ruled by varying methods and approaches. This vision of the world has one fundamental drawback: it is a vision. There is a world that underlies the layer of phenomena quantified and classified by experimental science and idealisms: this world we live in. And the world that is lived rather than observed and quantified is invisible. One of Henry's examples that best illustrates the limitations of this intellectual naiveté resonates with philematology, a type of knowledge that, as the previous essay argued, also unfolds in the dark. To these naïve thinkers of the world, Henry says, "the kiss lovers exchange is only a bombardment of microphysical particles."[9]

In his intellectual reservations against sight and in his privileging of the invisible as a ground particularly fertile for philosophical intuition, Henry's thought is in tune with that of the most prominent Epicureans. Lucretius had reminded his addressee Memnius that atoms cannot be grasped by sight because "nature works by means of bodies unseen," that is, bodies that touch.[10] In the Renaissance, Lucretius reappears after over a millennium of hibernation and causes a major stir. Well into the sixteenth century, the very term "Epicurean" was used as a pejorative epithet. According to explorer and naturalist Pierre Belon, during the siege of Bruges in 1562, Huguenots insulted Catholics calling them "Epicureans" (*Epicoriens!*), by which they meant "atheists" and "degenerates."[11] Epicureans were certainly no atheists, at least not in the contemporary sense of the word: they believed in the existence of the gods. But they did contest the notion of providence and the belief in divine intervention. And if their highest aspiration was a life of pleasure, this by no means implied hedonistic revelry and orgiastic frenzy—as many Christian thinkers since the early centuries of the Common Era adduced—but merely the absence of pain and bodily distractions. Epicureanism is, thus, neither atheism nor libertinism, but materialism. And rightly understood as such, it had a powerful impact on the haptic sensibility of early modern French culture.

In spite of the 1624 condemnation of atomism by the Sorbonne, the French received Epicurean philosophy and the works of Epicurus and Lucretius with considerably less apprehension than others in Europe at the time. The first French edition of *On the Nature of Things*, by Bade and Petit, with Denys Lambin's commentary, was published in 1563 and read, reread, and discussed by some of the sharpest minds of the time, most notably Michel de Montaigne, whose copy, showing extensive underlying and marginal notes, has been preserved.[12] The first translation of Lucretius into French, by Michel de Marolles, came out in 1650. By then, many atomistic ideas had already entered the bloodstream of French intellectual culture and this was, to a great extent, thanks to physician and philosopher Pierre de Gassendi, who some three decades before had launched what his biographer Bernard Rochot calls "the Epicurean project."[13] What began as a meticulous and enraptured reading of Diogenes Laertius's biography of Epicurus eventually became a philosophical life journey with deep roots in atomistic thought. Following in the footsteps of Fracastoro, Bruno, Bacon, and Galileo, Gassendi was instrumental in absorbing some of the key epistemological tenets of Epicureanism and adapting them into modern thought. Although Gassendi makes enormous speculative efforts to Christianize Epicurean thought, his "corpuscularism," like ancient atomism, is based upon an ontological and epistemological premise that posits infinitesimal bodies in contact with one another as the basis for generation, corruption, and knowledge of the world. Gassendi is among the first thinkers who attempt to close the Cartesian gap between *res cogitans* and *res extensa*, and he does so using atomistic tools.

The soul, he says, is not incorporeal, it is formed by corpuscles abysmally more subtle than those that form the body. This means that the difference between body and soul is not qualitative, but of mere material density.[14] In his main work, *Institutio Logica et Philosophia Epicuri Syntagma* (1658), Gassendi insists time and again on the infallibility of the sensorium, source of all our knowledge, and emphasizes the prevalence of the sense of touch. He then distinguishes three varieties of the tactile that prefigure the notion of the haptic. The first type refers to superficial contact between the body and anything in the outside world. The second bespeaks the sensation of discharge leaving the body. The third one is the sensation of the body as interiority, from the beating of the heart to the process of digestion.[15] In other words: exteroception, excretive sensation, and intraception. Despite his theoretical acrobatics to Christianize materialism, in emphasizing the complexity and the importance of the sense of touch Gassendi aligns himself with Lucretius and with Epicurean ideas that, if pushed enough, seriously contest the dogma of divine providence. However, the third-person detached and scholastic style of his philosophical prose differs greatly from Lucretius and Epicurus's second-person addresses. Whereas the pioneers of atomism make themselves present in their writings, Gassendi disappears behind the voice of an incorporeal *cogito*. Furthermore, unlike Condillac and Maine de Biran not too long after him, Gassendi neglects the issue of the body as origin of all experience, sensitivity, and inquiry. Perhaps, a more effective approach to the issue requires that one make oneself present, embodying and inhabiting the text.

Two years after Gassendi's death, a former student of his, Hercule-Savinien de Cyrano de Bergerac, published *Voyages to the Moon and the Sun* (1657), a novel in two parts that conflates science fiction with orthodox atomism. The work is a delightful first-person narrative and the narrator's presence in the text is all the more powerful by the emphasis the story places on the sensations and affections felt by him while in outer space. The premise of the voyage is also remarkable: the protagonist wishes to corroborate firsthand Copernicus and Kepler's astronomical theories, as well as the notion, proposed by Bruno, Galileo, and Gassendi, that the universe is infinite. Cyrano, however, understands infinity as a haptic phenomenon. "Infinity is simply a texture without bounds,"[16] the narrator says toward the beginning of the novel as he is preparing himself for the journey. What makes interspatial travel possible are precisely the materiality and homogeneity of the universe.

The main character arrives at the moon onboard a preposterous flying vehicle and finds a utopian world inhabited by intellectually exquisite beings. Lunar philosophy, as the narrator soon finds out, represents a violent transvaluation of Christian values. It denies the existence of God, rejects the notion of divine providence, does not believe in the immortality of the soul, and posits physical pleasure as the highest good. Lunar philosophers have a fair share of knowledge about our history of philosophy, and they despise Platonism and Aristotelianism with equal passion for considering them wild superstitions that cater to desperate, gullible souls. The protagonist introduces them to atomism, the noble school founded by Democritus, continued by Epicurus, and then by "the great poet" Lucretius. They listen in awe. As it turns out, atomism bears an uncanny resemblance to their conception of the world. Like Lucretius, lunar philosophers consider all senses to be variants of an overarching sensitivity that is tactile. The inhabitants of the moon, however, don't have five but infinite senses, that correspond with the infinite worlds. By means of this hyperbolic sensitivity they are able to perceive everything from changes in the ocean tides to empathetic vibrations between living beings. One of their philosophers tries to illustrate this hypersensitivity for the narrator: "If I tried to explain to you what I perceive through senses which you lack you would conceive it as something which can be heard, seen, touched, smelled, or tasted, when it is nothing of the kind."[17] We are limited to apprehending the world through the senses that we have, but these can be refined and pushed to unsuspected limits. In fact, from the moment he arrives at the moon, the protagonist develops new forms of tactile sensibility.

The word Cyrano uses to encompass this variety of novel sensations is *chatouillement* ("tickling," or "titillation"). As soon as he sets foot on the moon, the narrator feels "tickled" (*chatouillé*) with the pleasant pangs that, he assumes, the embryo must experience when the soul is infused in it.[18] The experience of being on the moon is fundamentally haptic, and the narrator makes strenuous efforts to account for an array of new sensations that he feels: tickles, titillations, vibrations, tremors, caresses. Given that all forms of pleasure are degrees of intensity in the tickling, one of the lunar philosophers criticizes the puritanical morals of terrestrial thinkers that condemn the most intense forms of tickling and hold certain parts of the body in higher esteem over others: "Is there any part of your body more sacred or more profane than another? Why should it be a sin when I touch my center-piece and not when I touch my ear or my heel? Is it because there is a tickling sensation?" And he concludes: your father "made you by tickling himself."[19] The moon philosopher rejects all forms of spiritualism and proposes a radical materialism. The infinite universe, every single ounce and corner of it, is tangible, and with a barely distinguishable vibration engages us through touch. But when touched, it also touches back, and herein lies its forcefulness. In Cyrano's, as in Lucretius's world, the intensity of the real is in its tangibility, and the power of the real is in its touch.

Among the most curious aspects of life on the moon there is the issue of language. There are two native languages, as the narrator soon comes to find out, and they are both fundamentally haptic. There is the language of the upper classes, which is musical and provokes a slight tickling on the hearer's ears, and that of the lower classes, a purely corporeal language that consists in making one's limbs tremble and shake.[20] Every movement and tremor of a finger, an arm, of the lips, of the hips means something. The intensity of movement shapes the meaning, and the combination of movements and rhythms articulates discourse. In order to communicate effectively, the moon dwellers from the lower classes must always go around naked and when they converse, the stupefied narrator points out, they don't look like people but like trembling bodies. The first form of language is apprehended through hearing, the second through sight, two faculties that operate by means of collisions between atoms. In adopting from Gassendi the Epicurean notion that all senses are variants of touch, Cyrano de Bergerac not only adheres to a new epistemology that regards the traditional hierarchy of the senses as obsolete but also anticipates what would soon become of the most pressing philosophical issues of the late seventeenth and eighteenth centuries: the problem of the relation between the senses.

"In France I frequented La Mothe Le Vayer and Gassendi. The second is a man who has written as much philosophy as the first has lived,"[21] says one of the moon philosophers reminiscing with the narrator about his trip to earth. François de La Mothe Le Vayer, along with Adrien de Monluc and Claude Le Petit, represent the trend in illustrated libertinism that rose to notoriety during the second half of the seventeenth century. In a letter dated 1653, La Mothe Le Vayer describes an encounter with a man who'd been born blind. The philosopher met the man in a dark room and asked him to go near the window to assess if he could tell darkness from light. The blind man admitted that, as he approached the dim light coming through the window, he perceived something different in the air, a certain density. He then added that this densification of the air was something that he often felt when he was near a wall, or when his face was close to an object; and he attributed it to "a kind of instinct . . . and to a pre-notion that nature gives him by means of the condensation of the air which he can somehow feel when he comes close to a physical mass that makes it rebound."[22] Thanks to this "instinct," similar to the echolocation by which bats orient themselves, the blind man perceives the proximity of an object and avoids bumping against it when he moves around the house. On the basis of this, La Mothe Le Vayer suggests that we pity and fear the blind because we are fixated with vision to the point that we believe that the highest enjoyments are all of the visual nature. Not only do we spend half of our lives in the dark, La Mothe Le Vayer notes, but, in order to indulge more fully in any form of sensorial pleasure that is not visual (a melody that softens our heart, a sip of wine, a kiss, a caress, or a fragrance that transports us to a cherished moment from the past), we normally close our eyes: "Darkness is sometimes so satisfying that we seek it out even in the most sacred things, for the dark air of temples increases our faith and the heavens never appear brighter to our souls than when we are in a dark place, or when it is night and we cannot see anything around us."[23] The notion of "dark air," with its synesthetic implications, refers back to the blind man's alleged instinct and suggests that said instinct is something that we all possess. Inside the temple, the worshipper sees very little but can perceive the darkness, and this perception is tactile. La Mothe Le Vayer concludes that sight, traditionally considered the most essential and precious of all senses, tends rather to be a source of distraction and error.

Almost a century later, in another philosophical letter, the case of the blind man who feels the light with his skin reappears, this time explicitly set against a background of Epicurean philosophy. The *Letter on the Blind for the Use of Those Who Can See* (1749) is also Denis Diderot's response to the problem of Molyneux, one of the most contested epistemological debates of the first half of the eighteenth century. In 1690, William Molyneux, an Irish scientist, had posed the following question: if a man who was born blind were to suddenly and miraculously recover his eyesight, would he be able to recognize through vision objects that he knew only by touch?[24] Unlike other philosophers, such as Locke, Hume, Condillac, Berkeley, Leibniz, Voltaire, and La Mettrie, who proposed answers, Diderot is less interested in the recovery of vision than in the first part of the problem: what kind of grasp on the world does a person born blind have? The Latin epigraph of Diderot's letter, "They can, but they don't seem to be able to" (*Possunt, nec posse videntur*), includes a pun involving the verb "to see" (*video*) that betrays a critical conception of sight. Like La Mothe Le Vayer before him, Diderot hopes to show that vision has been excessively overestimated in Western culture and that the blind, mainly by means of a hypersensitized tactility, are by all accounts self-sufficient and often even more sagacious than those who can see.

Diderot's letter features two examples of blindness. The first one is that of a man from Puiseaux, whom the *encyclopédiste* had met in person, whereas the second one is that of Dr. Nicholas Saunderson, a Newtonian mathematician and proponent of a tactile arithmetic. The blind know the world through touch, says Diderot, therefore the world to them is relief and texture. Sight, on the other hand, is a hampered form of touch. "Sight, so he is bound to conclude, is a kind of touch that only applies to objects other than our faces, which are located at a distance for us," Diderot argues.[25] When asked whether he'd like to be able to see, the blind man from Puiseaux replies: "If I wasn't so curious, I'd just as well have long arms as it seems to me that my hands could teach me more about what's happening on the moon than your eyes or telescopes can, and besides, eyes stop seeing well before hands stop touching."[26] Unlike Molyneux's hypothetical blind man, Diderot's subject, an actual individual, has no interest whatsoever in seeing; his touch provides him with sufficient experience. Similarly to La Mothe Le Vayer's blind man, the man of Puiseaux also possesses a haptic instinct that helps him "work out how . . . near he is to other bodies by the way the air feels on his face."[27] So too Dr. Saunderson had developed an epidermal hypersensitivity, adds Diderot later, to the point that he saw with his skin: "He had such an exquisitely sensitive epidermis that if a draughtsman were to sketch a friend's portrait on his hand he would undoubtedly have been able . . . to recognize it."[28]

Saunderson's epidermal hypersensitivity informs his vitriolic anticlericalism. The mathematician is the voice of Epicureanism in Diderot's letter, and the channel through which the author expresses his profoundly antiprovidentialist position. As the mathematician lies on his deathbed, he is visited by a priest who comes to perform the last rites for him. The holy man hopes to convince him that God, in fact, exists. The only logic the mathematician believes in, however, is that of nature understood in a very Lucretian vein as a constant cycle of trial and error. Some of Saunderson's responses to the priest's claims anticipate arguments later made famous by Lamarck and Darwin. The most Lucretian moment of the letter comes when Saunderson challenges the priest with the following proposal: "If you want me to believe in God you must make me touch him."[29] Diderot's conclusion, after his meetings with both men, is that "touch, with practice, can develop and become more sensitive than sight."[30]

Diderot's letter, like La Mothe Le Vayer's, and like Cyrano de Bergerac's novel, is a first-person narrative. The imaginary context of the moon allows Cyrano to evoke a vast variety of tactile experiences of unprecedented subtlety, as well as imagine a purely haptic language. By making himself present in the narrative, by describing the tickling he felt with his palpitating body and his titillating skin while on the lunar surface, the author produces a stark contrast between the credibility of the first-person testimony and the fiction of space travel. This interpretative dissonance functions as a caveat for the reader; sense perception is something utterly personal and nontransferable, Cyrano seems to be warning. Just as the narrator could not grasp the enormous variety of senses that moon dwellers possess other than by translating the description of them into data accessible to our five senses, also here on earth sensorial experience, with all its overwhelming expanse of connotations and nuances, is largely a private affair. And the language that describes sensation is approximate at best, imperfect, imprecise, and even misleading because description is, after all, a form of translation. In his treatise on beauty, Plotinus summarizes the esoteric principle of mystical experience with his famous "anyone who has seen It knows what I mean when I say that it is beautiful."[31] Cyrano de Bergerac would paraphrase, "Anyone who has touched knows what I mean."

To ponder the intricacies of touch is a first-person endeavor also for La Mothe Le Vayer and Diderot. Both men are aware that conveying the meaning of some of the things that the blind say poses considerable difficulties. The instinct that helps La Mothe Le Vayer's blind man to perceive the proximity of hard surfaces or the densification of the air felt by the blind man from Puiseaux is a baffling experience that we can barely comprehend, let alone experientially grasp. The blind are for Diderot and La Mothe Le Vayer ideal subjects on whom to study the relation between the senses and, in particular, the crucial role of touch because by being deprived of sight they are also exempt from its misleading distractions. Sight is, for both these thinkers, the most overestimated and the most pedestrian of all senses, while touch is the deepest, truest, and most philosophical.

From these early modern pioneers in the exploration of the frontiers between the senses, I now jump forward to the year 2000 when, not long before his death, Michel Henry published *Incarnation*, a work that summarizes over half a century of inquiries and represents a true philosophical testament. The work proposes a reversal of traditional phenomenology, but it is especially aimed at Merleau-Ponty, who dedicated particular attention to the problem of the body. In his critique of Merleau-Ponty, Henry rethinks the legacy of the different strands of empiricism and sensualism—or sensationalism—that proliferated in England and France between the seventeenth and the eighteenth centuries. With the exception of Maine de Biran, whom he considers a direct precursor of his phenomenology of the flesh, the great error of sensualists such as Condillac—later repeated by Merleau-Ponty—was, Henry argues, failing to recognize that there is an instance that precedes sensitivity. In his considering the chasm between "touching-touched" as degree zero of phenomenology, Merleau-Ponty misses the crucial point that this binomial is an epiphenomenon of the self-affection in which life reveals itself as flesh.[32] To touch something, even our own body, is always preceded by a "being able to touch," Henry's *je peux* ("I can").[33] In other words, for Henry, Merleau-Ponty is interested in how the body, understood as incarnated consciousness, accesses the world but neglects the issue of the body in and of itself. And this makes Merleau-Ponty fall back into the trap of intentionality; his approach invariably opens between the body and the world the old Cartesian gap that isolates one and alienates the other.[34]

By proposing an instance of radical immanence like self-affection that is grasped through what he calls "arch-intellegibility," Henry ventures into an uncharted territory refractory to all vision where distance doesn't exist. In the precorporeal and presensorial reality of the flesh, Henry finds not only self-affection but also the condition of possibility for movement and volition. Without using the term, Henry is mapping out an archeology of the haptic. Variants of the haptic phenomenon such as affect, movement, balance, proprioception, and the sense of touch in its three modalities (exteroception, intraception, and excretive sensation) find in Henry's invisible and immanent flesh their cornerstone. Whereas the example of the blind man approximated La Mothe Le Vayer and Diderot to liminal regions of sense perception and forms of synesthesia that the notion of the haptic will later claim for itself, Henry's concept of invisibility illustrates the presensorial and proto-haptic nature of life's most intimate core.

Up until this point I have managed to avoid broaching the most important point of *Incarnation*, namely the Christian dogma of the Verb made flesh. Henry finds in this notion the most direct and palpable anticipation of his own phenomenology of the flesh. The belief in God made flesh, that riddle wrapped up in an enigma, holds the key, says Henry, to that other great mystery, that of our own flesh, that is revealed in a presensorial self-affection. The incarnation of the Verb is its self-revelation, Henry adds, and God reveals Himself as life when He becomes flesh.[35] Similarly, self-affection is a form of self-revelation by which one grasps oneself as life and as possibility of movement. A phenomenology of the flesh thus becomes a phenomenology of incarnation. Against the grain of mainstream Western philosophy—a conglomeration of traditions blinded by the light of ideas and fixated with clarity—Henry finds in Christianity, especially in some of the early Fathers of the Church, a profound and committed attempt to engage with the problem of the flesh and the body in the pitch-black darkness. This might sound like a paradox; after all, over the last two millennia Christianity has also produced blasting condemnations of the body for being an ever-open gateway to the temptations of the flesh, it has instigated fear of the passions, and warned against the perils of tactile enjoyment. But didn't Lucretius, the grand champion of touch, also advise against the evils of bodily excess and the distraction that the sense of touch (as vehicle of both pleasure and pain) causes? Making the body, even the flesh, the center of philosophical inquiry by no means implies a tacit condonement of all its whims and desires. On this point, Christianity and Epicureanism agree. Not surprisingly, one of the most fascinating examples of this unlikely accordance is to be found in a French author.

For Christians, the flip side of incarnation is the Eucharist.

God is made flesh to save mankind, and mankind, in turn, must eat God's flesh to be saved. "He that eateth my flesh, and drinketh my blood, dwelleth in me and I in him," reads John 6:56. Toward the end of the seventeenth century, the bishop of Meaux, Jacques Bénigne Bossuet, references these words in one of his sermons and illustrates them by hinting at the famous lines of book 4 of *On the Nature of Things* where Lucretius explains sexual desire by dissecting the phenomenon of the kiss. Euphoric, Lucretian lovers open their mouths, salivate, and bite each other in their vain struggle to possess, or to become one with, one another. For Lucretius such attempts lead to nothing because the atoms that compose us are impenetrable and immutable. In front of an audience that was clearly familiar with these raunchy, pagan verses, Bossuet explains the Eucharist as a ceremony that breaks through the dead-end road of materialism and makes the union of two in one possible:

> In the ecstasy of human love, who is unaware that we eat and devour each other, that we long to become part of each other in every way, and, as the poet said, to carry off even with our teeth the thing we love in order to possess it, feed upon it, become one with it, live on it? That which is frenzy, that which is impotence in corporeal love is truth, is wisdom in the love of Jesus: "Take, eat, this is my body": devour, swallow up not a part, not a piece, but the whole.[36]

The bishop's words depict with striking voluptuousness the intrinsic carnality of the Christian dogma that Henry often invokes. But they also underscore this unsuspected affinity between the atomistic and the Christian sensibilities. Perhaps, such an affinity could only be possible in France, where Epicureanism spawned sensationalisms both chaste and libertine, moderate and erudite, and where phenomenology attempted to reach the farthest depths of the human experience, the very mystery of the living flesh that manifests itself like tremor in the dark night of the soul. But if such an encounter was possible it surely means that there was something preceding it and enabling it, something transhistorical and primal, a constitutive element in the very fabric of the French cultural sensibility—something undoubtedly ineffable and ungraspable that, accustomed to images as I am, I cannot help but to envision as an ever trembling body, or a layer of skin covered in goose bumps.

How can man know himself? He is a thing dark and veiled;
And if the hare has seven skins, man can slough off seventy
Times seven and still not be able to say:
"This is really you, this is no longer outer shell."
—Friedrich Nietzsche (1874)[1]

SKIN DEEP

Almost four centuries ago Sir Thomas Browne wrote: "As though there were a Metempsychosis, and the soul of one man passed into another, opinions do find after certain revolutions, men and minds like those who first begat them."[2] This statement has the remarkable quality that as soon as one agrees with it, it automatically becomes a proven fact. But unlike the souls that patiently await future incarnations by the shores of the Lethe, for Browne opinions "find" the minds and people (bodies) they later inhabit. And if they find them it is surely because they seek them. So we can picture throngs of opinions, ideas, and—I may add—literary images, artistic motifs, aesthetic forms sniffing around the vast fields of history, vehement like roving hounds searching for receptive minds and bodies compatible with them. Whereas, according to the ancients, souls are never purer and nobler than once they are free of the body, Browne's antic posits ideas and aesthetic forms only being able to flourish *within* a living body. Opinions and ideas are thus essentially connected with their specific embodiments. And, as they find bodies to inhabit in the course of their relentless search, they also connect minds through the centuries—minds that channel them, that shape them, and that are shaped by them in turn. In this way, men and women who have lived tens, hundreds, thousands of years apart form invisible chains that are lost in the distant past and that extend into the uncertain future. Browne's strange notion conjures up a vivid picture of transhistoricity and suggests mental but also corporeal affinities that defy specific cultural (and, therefore, ethnic, gender, social) identities. Later in the text he adds: "Men are lived over again, the world is now as it was in ages past."[3] But if such transmigrations happen from individual to individual they are also surely enacted at a collective level; and just as there are compatibilities between minds and bodies throughout history, there are also affinities between times, epochs, eras.

It is the dawn of the twentieth century and Hugo von Hofmannstahl is writing a letter. He is not yet thirty, but his peers already consider him a prodigy, the greatest poet of his generation. The letter he is penning, and that will be published in the Berliner newspaper *Der Tag* on October 18 and 19 of 1902, is not just a break from poetry. It is a manifesto; the manifesto of a cultural sensibility that arises in the late nineteenth century, that will dominate the first decades of the twentieth century, and that will revolutionize forever the ways in which we think about, feel about, and go about producing art and literature. It is a most peculiar letter, though: addressed to Francis Bacon, dated August 22, 1603, and signed by one Philipp Lord Chandos, poet and youngest son of the Earl of Bath. Hofmannstahl's letter came to be so influential that in German it is known simply as *Ein Brief* ("A Letter"). In it, the author makes an urgent announcement regarding a collective cultural crisis (perhaps irreversible) of which the fictitious and melancholy Lord Chandos first becomes aware one day as he peruses with extreme attention the geography of his hand.

Almost exactly three centuries before the young Viennese poet sat down to write the Chandos letter, his fictional addressee, Francis Bacon, published one of the most fundamental works of early modernity. *The Advancement of Learning* (1605) both celebrates and paves the way for the scientific revolution that swept through the West between the sixteenth and seventeenth centuries. Bacon invites scientists and natural philosophers to temper their reverence to book learning and concentrate on the direct observation of nature and on firsthand experience. One of the most powerful enemies of scientific advancement, for Bacon, is language with its empty jargons and its obsolete concepts inherited from past traditions. In the *Novum Organum* he will famously call such notions "the idols of the marketplace." For the young Lord Chandos in Hofmannstahl's letter, language is also an overwhelming nuisance; in particular, he condemns the naïve belief that words and concepts are capable of accurately representing the world. At one point, the poet confesses having completely lost "the ability to think or speak coherently about anything at all."[4] Words fall apart in his mouth like rotten mushrooms, and nothing is more challenging than mindless conversation, gossip, opinion, chitchat: "My mind forced me to see everything that came up in these conversations as terrifyingly close to me. Once I saw through a magnifying glass that an area of skin on my little finger looked like an open field with furrows and hollows. That's how it was for me now with people and their affairs."[5] This rejection of conventional language that forces the alienated young man to suspend all social and literary activity directs him to concentrate on the surrounding world. He no longer finds pleasure in words and in rhyme, says Lord Chandos, but in ineffable epiphanies that he experiences upon seeing, for example, a dog lying in the sun, or a homeless man in the street, or a harrow forgotten in a field. The most brilliant accomplishment of the letter lies in the stark, yet subtle, contrast it produces between the denunciation of language as bankrupt and unable to convey meaning, and the crystal clear, exquisite prose in which it is written. The Renaissance decorum, the sober tone, and the perspicuous language filled with sublime images produce a striking dissonance with the alleged incapacity to express anything coherently. There is a pronounced incongruity between form and content, and form prevails. The image of the abandoned harrow refers back to Lord Chandos's little finger seen through the magnifying glass with its furrows and hollows, like an open field of skin, and suggests a need to hold on to form. From a distance, concepts and words crumble down and the poet, in a desperate attempt to avoid being sucked into the chaotic whirlpool of nonsense, comes closer and closer to things, to objects, to his own body, until sight becomes touch and he can hold on to something. Through the magnifying glass he traverses the landscape of his own body and finds nooks hitherto unknown; his eyes operate like a harrow that plows this *terra incognita*, restoring sense to it and appropriating it by means of the artifice.

The *Letter of Lord Chandos* is a manifesto of modernism. It calls for a new foundation of language and it proposes for the task a new kind of poetics that, instead of striving to reflect the world, aspires to create new, artificial, parallel worlds that merely brush against material reality (whatever that may be) like an eye brushing against unsuspected textures through a magnifying glass. Hofmannstahl's call is an appeal to remain on the surface and unravel its nuances. A late aphorism, included in his *Book of Friends* (1922), illustrates this with mysterious poignancy: "It is depth that we must hide. But where? On the surface."[6] Although modernists were fond of the paradoxical and the aporetic—the more provocative, the better—Hofmannstahl's appeal is far from being a mere trope. Instead, the Austrian writer seems to be announcing an exhilarating discovery. Perhaps that old Western belief that locates the essence of what we are in the inside of the body, inaccessible to the senses, intangible, or outside in a sphere of transcendence, is false. Perhaps our nature, our "essence" (if there is such a thing), lies on the very surface. And the surface for us is the skin.

The skin is the most vast and complex organ in the human body. It is composed of connective tissue, blood vessels, lymphatic nerves, elastic fibers, special cells; and it is separated in two large regions, the epidermis and the dermis, divided by the so-called basement membrane. Whereas the epidermis has four to five different layers, the dermis consists of two regions, the papillary and the reticular. In an average adult weighing around 160 pounds, the skin has an extension of about twenty-two square feet and weighs around seven pounds. Think of a twin-size blanket; displayed in all its glory your skin is probably larger and heavier. The skin is the unsurpassable frontier that at once separates us from the world and connects us with it. It is the wrapping that protects our inner organs—too fragile for the outside world—and also the organ of exteroception, which allows us to feel textures, to experience temperature, pleasure, and pain. Homeostasis, or the regulation of bodily temperature, is one of its most important functions, and its vexing efficiency suggests the existence of a certain dermic intelligence. So too, the skin's ability to regenerate is one of the most astonishing qualities of the human body.

The physiological importance of the skin is staggering, but its cultural significance is no less crucial. Skin constitutes the very foundation of individual and collective identity, halfway between nature and culture, and its study stands on the threshold that divides the natural sciences from art and the humanities. Of all its characteristics, perhaps color is the one that best expresses the skin's role as intermediary between nature and culture. The association between skin and pigmentation has a long history that transpires from very early on in the Indo-European languages. The Latin word for color (*color*) comes from a Sanskrit term that refers to milk skin. The Greek word *chroma* (color) shares the same root with *chros* (skin) and *chrozein* (to touch a surface). Beyond its natural color, which can be modified, lightened, darkened, the skin is a space for decoration. Tattoos, piercings, plastic surgery, body paint, makeup, creams, and cosmetics in general are common marks of personal and collective identity. But the skin also functions as a reliquary of past experiences; it is the living book on which the history of our lives is written and it acts as the incarnated memory of the body. Two scars on my right hand and one on my chin tell the story of old accidents of which I have little to no recollection. A constellation of freckles on my shoulder blades is the alarming souvenir of a New Year's morning hangover lying under the blazing South American sun. And a collection of incipient, yet slowly expanding wrinkles on my face marks the passing of time with irrevocable authority. Our skin is a canvas on which two sets of hands (ours and the world's) paint the picture of our lives. It is the palimpsest that tells our story. It is the ragged and cherished flag under which we sail toward death, and it is our first shroud.

One of the most interesting contemporary studies of the ways in which skin molds our psyche is Didier Anzieu's *The Skin Ego*. The work centers on the notion of *moi-peau* ("I Skin"), introduced in the 1970s, by which the French psychoanalyst meant: "a mental image of which the ego of the child makes use during the early phases of its development to represent itself as an ego containing physical contents, on the basis of its experience of the surface of the body."[7] Anzieu's concept builds on less-explored intuitions and speculations regarding the relationship between the psyche and the surface of the body dating back to the earliest days of psychoanalysis. Freud's writings on sexuality explore precisely the close relationship between the unconscious and the erogenous zones of the body. In his study on sexual aberrations, the first of the *Three Essays on the Theory of Sexuality* (1905), Freud underscores the libidinal importance of touch and sight (a sense that, according to him, is derived from touch) and points to the skin as erogenous zone *par excellence*. The concept of "instinct" (*Trieb*) that articulates Freud's sexual theory is particularly revealing as it refers to the frontier between the physical and the mental. He concludes that skin, understood as a second frontier, is the main vehicle for instincts to be satisfied. Freud began his career as a physician in Vienna's General Hospital, barely a year after the death of Joseph Skoda, and not much longer after those of Karel Rokytansky and Ferdinand von Hebra, founding fathers of Vienna's School of Dermatology. During at least four years, he worked just a few feet away from Heinrich Auspitz, the heir of this tradition. Freud's investigations, as much as those conducted by the pioneers of modern dermatology, are the token of a profound interest on surfaces that emerges during the last decades of the nineteenth century and that intensifies and expands well into the twentieth century. This fascination with superficiality, with texture, and with touch, manifested most notably in Aloïs Riegl's (who was also Viennese) 1902 coining of the term haptic, also ignited the imagination of many artists and writers.

But if, as Thomas Browne argued, opinions transmigrate always in search of new minds and bodies, then every discovery must be a rediscovery. And just like the nomads who crossed the Bering Strait came upon the new world ten thousand years before Columbus; just like Anaximander and Lucretius intuited two millennia before Darwin that species evolve, early modern physicians and artists had discovered the crucial importance of the skin and the fathomless depth of surfaces, centuries in advance. Now, given that we are talking about skin, why not unfurl an anatomical banner? We're in Oxford. The year is 1629, or 1630. It is winter—dissection season—and from a window of the first floor at the southern end of the Bodleian courtyard hangs the tanned skin of a man. Seven pounds, twenty-two square feet. It is a shocking sight, even for medical students used to anatomical spectacles and severed body parts, who walk up and down the courtyard on their way to and from class—even for Thomas Browne, who is about to graduate and move to the Continent for three years, where he will study in some of the best medical schools of the time. Over forty years later, the then old Norwich physician recalls the daunting vision of the flayed skin in a letter to his son: "To flay men dead and tan their skins is not now a practice, but above forty years ago, the tanned skin of a man was hung up in the physics school at Oxford."[8] Leaving aside the morbid curiosity that such a sight might produce in an onlooker who is thirsty for strong emotions, or just hungry for scientific insights, I suspect that the spectacle was memorable for Browne also because of its symbolism. The tanned skin of that man, waving in the cold northern wind, was a flag that announced and celebrated the discovery (or rediscovery) of the human body.

What came to be known as the "anatomical revolution" is a process that began toward the end of the Middle Ages and that is directly related to the gradual acceptance and standardization of the practice of dissecting human cadavers for medical research. It reaches its peak in the mid- to late sixteenth century with the works of Vesalius and Colombo, Estienne, Paré, D'Acquapendente, Harvey, and others. For practical reasons, dissections began with the inner organs, which are the first to decompose. This has led some critics, like Claudia Benthien and François Dagognet, to interpret the history of anatomy as a "reversed archaeology" that privileges the interior of the body and discards or neglects the surface.[9] Nothing could be farther from the truth. In fact, far from discarding the surface, during their exploration of the deepest and darkest cavities and crevices, early modern anatomists discovered the skin.

The schools of medicine founded in antiquity by Hippocrates and Galen, which dominated medical studies until well into the modern period, considered skin a mere wrapping or covering of the body that completes its structure and that, thanks to its conveniently located orifices, allows for the evacuation of excretions, thus ensuring a healthy balance. Any abnormalities on the skin's surface were considered mere cosmetic issues: deformities, imperfections, and irregularities, rather than diseases. It is during the Renaissance that this begins to change. As they started separating the different layers that form the skin and realizing the immense complexity of the integumentary system, anatomists soon understood its organic importance. In times when physicians wrote epic poems, natural philosophers composed dialogues, and the now seemingly irreparable gap between the natural and the human sciences had not yet opened, the news of the discovery of the skin ran like wild fire, influencing artists, intellectuals, and thinkers of all backgrounds and affiliations. Poets, playwrights, philosophers, theologians, painters, sculptors express through their works the enthusiasm caused by the discovery (or rediscovery) of the most fascinating and unknown continent of all: the human body.

Artists were among the biggest enthusiasts of this new anatomy. Already in the fifteenth century, men like Leon Battista Alberti and Lorenzo Ghiberti advised painters and sculptors to study human anatomy in order to perfect their figurative technique. We know that Pollaiuolo, Verrocchio, Michelangelo, Leonardo Da Vinci, and others had not only witnessed but even performed dissections, cutting, removing, manipulating, opening, and flaying with their own hands. Pollaiuolo's *Battle of the Nudes* (ca. 1470) is often identified as one of the earliest and clearest examples of the intersections between the visual arts and the anatomical revolution. Some of the bodies depicted in the engraving seem to have been completely flayed, their muscles and veins exposed, the flesh raw. Perhaps the most famous iconographical example of a skinless body is Michelangelo's Saint Bartholomew. The picture of the martyr, who according to tradition was flayed alive and crucified upside down in Armenia, appears at the center of the *Last Judgment* (1536–41). The saint is holding a knife with the right hand whereas from the left hangs the bundle of his whole skin. It has been argued that the crumbled face on the loose, hanging skin is Michelangelo's self-portrait. Whereas Pollaiuolo's bodies looked like they had been flayed, the only sign we have that Bartholomew has suffered the same punishment is the skin that he holds. The apostle's body, perfectly muscular and colorful, is not the martyrized and gaping wound that the historic Bartholomew certainly was after undergoing such torment. This refers to the fact that, according to Christian belief, at the hour of the resurrection of the flesh the body that rises to meet its Maker is not the rotten, mummified, fragmented corpse resting in the grave, but the person at the height of their life. Although perhaps Michelangelo, a well-seasoned anatomist, was also revealing in an oblique way something that he had corroborated firsthand at the dissection table: layers hide more layers, and under the skin there is just more skin.

Leonardo's interest in anatomy is even more notorious because, throughout his life, the artist compiled a vast collection of illustrations of the human body with the intention of producing a manual. The project was interrupted by his death in 1519, but it was recently undertaken once again with a critical edition of all the remaining illustrations and an exhibit at the Palace of Hollyroodhouse, in Edinburgh.[10] Few understood the importance of anatomy in Leonardo's work better than Paul Valéry. In the notes and digressions that accompany his *Introduction to the Method of Leonardo da Vinci* (1894), the French poet says: "This man who dissected ten cadavers to follow the course of a few veins thinks to himself: 'The organization of our body is such a marvelous thing that the soul, although something divine, is deeply grieved at being separated from the body that was its home.'"[11] This love of bodies, this veneration for sensible shapes, this attention to the detail of surfaces that borders on devotion is what makes Leonardo stand out as the great modern genius, according to Valéry. The poet finds in Leonardo's art a secular liturgy of sorts in which he sees a direct antecedent of his own aesthetics—a type of aesthetics both formally orthodox and unashamedly sensualist. Years later, in *L'idée fixe* (1932), a philosophical dialogue that has the very Socratic objective of exposing the falsity of a number of common places and vulgar opinions, Valéry refers to the issue of depth. When one of the interlocutors confesses his extreme dislike of the notion of "depth" applied to feelings, people, matters, the following exchange ensues:

> — I remember what the medical books say about the development of the embryo. One fine day, it creates a fold, a crease in its outer envelope . . .
> — The ectoderm. And it closes up . . .
> — Alas, yes! . . . And all our troubles start from there! . . . Chorda dorsalis! And then marrow, brain, everything needed to make us feel, suffer, think . . . and be profound. That's where it all starts.
> — So what?
> — Well, the whole process is invented by the skin! . . . However much we dig, doctor, we remain . . . ectodermal. . . . Nothing is deeper in man than his skin—in so far as he knows himself.[12]

There is nothing deeper in us than skin because it all begins as a random, spontaneous fold in the embryo. Every movement, every feeling, every thought, every word is born from that primal contortion in the ectodermal membrane. While Valéry's characters discover (or rediscover) the depth of the surface through notions of embryology, early modern anatomists understood this as they patiently and skillfully removed the skin off of cadavers. As early modern anatomists flay away, ever more meticulously, they discover that layers have layers of their own. And for a moment, much like a child opening a Matryoshka doll, they stop thinking of what treasure lies hidden in the inside and just looks forward to the next layer. This leads to a change in the way the body as a whole is conceived. Skin continues to be a cover, or envelope, but it no longer is a *mere* cover. As the focus turns to the surface, two things happen: on the one hand, the notion that skin is a sensitive organ progressively gains more advocates. Touch is diffused throughout the body, through the skin, especially in the hands, whose *cuticula* is the most sensitive. On the other hand, this growing concern for the physiology of the skin leads to the first systematic treatise on diseases of the skin, *De morbis cutaneis* (On the Diseases of the Skin), by Girolamo Mercuriale, in 1572. In this treatise, the author emphasizes the way in which skin—rather than simply being a cover for the inner body, and because it is connected to the inner body by veins, nerves, and tissue, is itself susceptible to pathology. Far from being a "mere covering," the nature of skin is organic, and I mean this in two senses: it is in itself a large, multilayered sensory organ, but it is also a vital part of a larger organism: the body.

The border between exterior and interior fades away as medicine begins to understand that the body is an organic, continuous whole. And as the skin starts to be recognized as a vital organ, the realization of its complexity opens the door to new ways of understanding the relationship not only between exterior and interior but also between body and soul. As they remove layer after layer of skin, anatomists grasp the unsuspected depth of its apparent superficiality, but, more importantly, they are struck by a revelation that is both awe-inspiring and disconcerting. Perhaps the ancient belief that distinguishes between an interior—invisible, intangible, immortal, essential, the core of who we are—and a corporeal, transient, superficial exteriority is nothing more than an illusion. Perhaps the atomists—whom, ever since Aristotle, many in the West derided and demonized systematically—were right all along. Perhaps in the darkest, most secret depths of our body there is nothing but more body.

In January 1921, Filippo Tommaso Marinetti published in the journal *Comoedia* a text that he had already read publicly in Geneva and Paris. Not unlike other writings by the father of futurism, this too was a manifesto. In it, Marinetti announces the invention of a new science: tactilism. *Il tattilismo* was revealed to him, he says, during a summer holiday on a beach in Antignano, not far from the Avenue Amerigo Vespucci.[13] This is not a coincidence; in fact Marinetti sees himself as a new Vespucci faced with an uncharted and unnamed continent: the skin. And just like Vespucci, who was not the first European to set foot in the new world, but the first who realized it was a new world, Marinetti announces the discovery of a science based on the most common and universal of all senses. Since the end of World War I, he argues, Europe has been divided into two groups of people: the majority, which strives for material progress, and the minority (poets, artists, thinkers), which wishes to leave the city and find refuge from the industrial world in the country. Futurism declares itself an enemy to both. The majority lives a life of superficiality and naïveté. The minority is sick with arrogance and resentment. A return to a bucolic, pre-industrial state is impossible. One must not only accept but also advance and even celebrate progress. Not, however, as a mere vehicle for acquiring wealth; instead, as the springboard for a revolution based upon love and friendship. This new man, says Marinetti, must privilege touch above all the other senses. The revolution is, in fact, a tactile undertaking.

Handshakes, kisses, hugs, and sexual intercourse shall be the only viable and significant forms of communication and thought exchange, announces Marinetti. Under tactilism, the sole means of communication is the epidermis and, in order to develop its potential to the fullest, Marinetti proposes a tactile education. He subsequently proposes providing a series of charts with different textures and materials, which will be used by hands and skin to train themselves in recognizing objects without seeing. These charts include elements like glass, silver, velvet, wool, silk, fish scales, horse fur, human skin, hair, iron, sponges, etcetera. In order to train our epidermis, Marinetti recommends exercises like swimming naked and trying to distinguish the different underwater currents by their temperature, or training ourselves in recognizing all of the objects in a dark room through touch.[14]

A year after the invention of tactilism and without any signs of the tactile revolution taking off, Benito Mussolini was appointed as Italy's prime minister. Marinetti and other futurists would soon become adherents and defenders of Fascism, a movement that they believed reflected many of their intellectual ideals. Marinetti's ludicrous and chimeric revolution, just like Valéry's epiphanies and Hofmannstahl's aesthetic insights, are isolated outbursts of enthusiasm in the midst of what is perceived by many as a desert of existential and cultural paralysis. In Marinetti's view, the enemies are Communist utopianism and capitalistic frenzy. For Valéry and Hofmannstahl, it is the commonplace, the cliché, and other naïvetés of language—self-complacent vices that further open the fissure between language and the world of things and feelings. The call to remain on the surface is sign of a haptic urgency, a need to hold on to something tangible, something that the first phenomenologists, around the same time, summarized in the motto "Back to the things themselves!" And the skin, border between an ever more disjointed interiority and a world that becomes progressively more absurd and ungraspable, is perceived in these first decades of the twentieth century as a highly symbolic space and as a most fertile ground to foster creativity.

The border is, in fact, one of the most common metaphors for the skin. It is of course a geographic metaphor, but it is also a political and, even, a metaphysical one. Be borders natural or artificial, they at once separate and connect regions. Given that it is by nature conventional, artificial, the very notion of border is tentative and *essentially* metaphoric. The skin constitutes the border between the inner and the outside part of the body, but also between the biological body and the body as a cultural construct, between the individual and the collective body, and between our own body and other bodies. Furthermore, skin is what makes of a body a *person*. When I use this term, which comes from the Latin *persona* ("mask"), I do it literally. Skin is the organic membrane that separates the body, understood as living flesh, from the person that we are. Skin masks us as persons, and like every mask, it at once hides and reveals. Vesalius remembers that when he was a professor at Padua—the mecca for early modern anatomy—he once found himself in dire need of a female cadaver to dissect. Given the custom of only anatomizing the bodies of executed criminals, acquiring specimens to work with was extremely difficult, and it was not uncommon for anatomists to resort to body snatching. Vesalius indeed sent a group of students to steal the body of a recently diseased woman from the local parish, and in order to conceal her identity lest someone recognize her, he had the body completely flayed before the dissection at the anatomy theater. The stratagem was a success. A body without skin is a body unmasked, depersonified, and it exposes the following paradox: the secret identity hidden underneath the epidermal mask is neither the trembling flesh nor the clammy organs, and not even the soul or the spirit; it is the skin itself.

A discipline that gained immense popularity in the Renaissance while anatomists were unraveling the mysteries of skin was physiognomy. Physiognomy studies the relation between physical appearance—in particular facial features—and temperament; in other words, the relation between what is tangible and what is intangible in us, our surface and our interior. Its origins are ancient, but during the Renaissance it gains unprecedented importance and popularity to a large extent due to the work of Neapolitan polymath Giambattista Della Porta. Della Porta wrote a physiognomical tetralogy that aspires to not only exhaust all relations between facial features and temperaments but also to draw a network of interconnections between human beings and animals, plants, and even celestial bodies based on their physical features.[15] The physiognomist's method is thus based upon direct observation and the close perusal of textures. Della Porta spent years doing field work at jails and insane asylums, studying, describing, and taxonomizing the hands and faces of mental patients and convicts. Since physiognomy is a science of textures, the physiognomist's eye must work like a magnifying glass brushing against surfaces. In the case of humans, appearance is a multifarious semiotic system that encompasses all the geography of the skin and includes hair, freckles and moles, wrinkles and dimples, bumps and protuberances, birthmarks and sunspots, color, texture, scars and any other distinctive marks. As a total system of sense, appearance lies before the skin as an object of study and, thus, physiognomy precedes anatomy. One could even say that appearance is the skin of the skin, a first layer too subtle for the anatomist's knife and only accessible with a haptic gaze.

Physiognomical knowledge turns the dichotomy surface-interior on its head. Whereas the surface of the body was understood traditionally as a veil, or a wrapping, that hides a truer reality lying underneath or beyond the reach of the senses, for the physiognomist that surface and that interiority are two sides of the same coin, two instances perfectly permeable to one another. Appearance, that phenomenon that manifests itself before the skin, but also *on* the skin, is no longer a curtain that one must open in order to access the true identity. It is, rather, the necessary counter side of that interiority and the most peculiar mark of singularity in any individual. The history of physiognomy does not end in the Renaissance. Every few generations this quirky pseudo-science resurfaces, like a prehistoric creature in a fresh-water lake, always inspiring new enthusiasts. Toward the end of the eighteenth century, and through the work of Swiss virtuoso Johann Kaspar Lavater, physiognomy experiences another renaissance. Lavater, an avid reader of Della Porta and of Sir Thomas Browne, breathes new life into the discipline with his *Physiognomical Sketches* (1775–78), the astonishing illustrations of which fascinated, among many others, Goethe. A century later, in 1876, Cesare Lombroso publishes *Criminal Man* where he advances his notorious thesis concerning the direct relationship between certain physical features and a propensity to criminality. Like Della Porta, Lombroso too conducted research at jailhouses and mental institutions, examining faces, scars, tattoos, and hands; and his criminological physiognomy was the object of considerable attention in Europe until well into the early twentieth century. Hans Gross, an Austrian criminal jurist and one of the fathers of modern forensic science, was one of many to engage with Lombroso's ideas in the German-speaking world. Although he often found himself disagreeing with Lombroso's sweeping generalizations, Gross was convinced that there was a direct correlation between a predisposition to crime and certain physical features. Between 1902 and 1905, he taught criminal law at the University of Prague. One of his students was Franz Kafka.

It has been argued that the random arrest at the beginning of *The Trial* is a veiled reference to Lombroso's ideas and to Hans Gross's classes in Prague.[16] The idea becomes even more suggestive when connected with Kafka's recurring interest in physiognomy, in skin, and in faces—an interest evident throughout his work, but that shines with particular radiance in his diaries. On October 13, 1911, he writes: "Inaesthetic transition from the taut skin of my boss' bald spot to the delicate wrinkles of his forehead."[17] A week later, on October 22 he notices "the striking smoothness of Mrs. Tschissik's cheeks alongside her muscular mouth."[18] And on December 16 of the same year he evokes "the warm shadow in the soft red of Mrs. Klug's mouth when she sings."[19] Kafka's narrative prose moves through the shallows, hovering over the folds and the bumps, the twists and the turns of bodies and things, and exposing, as if by means of a microscope, the horror, the grotesque, and the absurdity of life seen up close. Once again we summon Lord Chandos. In a letter to Adorno, Walter Benjamin suggests that Kafka took over Hofmannstahl's unfinished project.[20] The suggestion is enticing and resonates especially with the fixation both authors had with the outward shell of things. Kafka's literature moves outside of people and things, barely brushing against them, combing the desert of existence and occasionally exposing, through the narrow cracks in the exterior of what is perceptible, an imperceptible dimension that is dark and overwhelming, and that will eventually devour everything like a black hole. In short, few things matter more to Kafka than surfaces. After all, isn't *The Metamorphosis* the chronicle of a dramatic change of envelope? Just as in *The Trial* we never find out why Joseph K. was arrested, *The Metamorphosis* never discloses why Gregor Samsa underwent his uncanny transformation. This reticence to revealing causes and reasons—the mark of the great storyteller, according to Benjamin—forces the reader to concentrate on the only thing that is offered: the appearance of the real, its wrapping, its clothes.

On March 12, 1912, Franz Kafka writes in his diary: "In the trolley rapidly passing by there sat in a corner, his cheek against the window, his left arm stretched along the back of the seat, a young man with an unbuttoned overcoat billowing around him. . . . Today he has become engaged and he can think of nothing else. . . . There was no real connection between him and the trolley. . . . Only the billowing coat remains [of him], everything else is made up."[21] All we know about this man, all that he *is*—at once overwhelmed and absentminded, perhaps slightly anguished, and shockingly disconnected from everything around him—is in the gentle flapping of his unbuttoned overcoat. This preoccupation with clothing and the idea that garments are the most essential aspect of an individual inspires the brief text entitled "Clothing," in which the wrinkles, stains, and folds of clothes are an allegory of the wear and tear of the body and the skin. But the most compelling example of this interest in surfaces is, without a doubt, *In the Penal Colony*.

This nightmarish fable tells the story of an explorer who arrives in a remote, deserted corner of the world to visit a penal colony in which the brutality of the living conditions is only matched by the sophistication of the punitive methods. The excruciating description of that "remarkable piece of apparatus" used to carry out executions occupies two-thirds of the story.[22] The narrator directs our attention to every detail, nook, and cranny of the ghastly apparatus that, by means of hundreds of needles that come out of a harrow-like device, inscribes the crime on the skin of the prisoner. The needles, like those of tattooing machines, perforate the skin and go deeper and deeper. The machine works on the body for about twelve hours, pricking, inscribing, horribly stigmatizing the skin. After six hours of torment, the prisoner—until then, ignorant of the crime for which he was convicted—begins to decipher the inscription in a tactile manner, through his wounds. The moment he makes sense of what he is being punished for, the prisoner learns his lesson, dies, and is buried in a mass grave. This is, if it all goes well. But Kafka is telling the story of the decline of this form of execution. The moment it is turned on, it becomes clear that the machine is not working properly, that it is not writing anything; there is no message, just punctures. The skin is plagued with senseless scribbling—death wounds with no redeeming quality. As in the letter of Lord Chandos, here too all meaning has been lost and language has gone bankrupt, a senseless and barbaric gibberish that cannot account for either the outside world or the sphere of inner life. Like Hofmannstahl in his fictional epistle, Kafka also presents the reader with the image of a useless harrow and stages the crisis of meaning over the geography of the skin. And the surface of things, shining like a massive gaping wound, remains the only region where one can find whatever little sense remains.

I go back to Oxford, the winter of 1629–30, the tanned skin of a man hanging from a window, and Thomas Browne observing it from the Bodleian courtyard. Some fifteen years later, in his second published work, the *Pseudodoxia Epidemica* (1646), Browne dedicates three chapters to the issue of pigmentation, where he affirms that nothing separates human beings more than the skin. He is referring, of course, to skin color, a matter he considered to be among the biggest mysteries of science. But he is also talking about physiognomy. A great admirer of Della Porta, Browne once said: "There are mystically in our faces certain characters which carry in them the motto of our souls."[23] Following Della Porta, Browne also adds that this is not a peculiarity of human beings only. "The finger of God has left an inscription upon all his works," said Browne, who liked to imagine God in the form of a hand that creates and heals, and divine creation as the monumental work of a pious and industrious artisan.[24] Like all artists, God too leaves his signature on his artifacts, and his signature has the form of texture, a geometric pattern one can find in all of reality. Browne discovers this universal texture in one of Della Porta's most obscure works, a treatise on gardening.

Villae, published by Giambattista Della Porta in 1592, is a treatise that discusses how to design, plant, and maintain gardens. In the fourth book, the author broaches the topic of the *quincunx*, or lozenge, and the *quincuncial* order, an archaic technique of planting trees in the shape of the number five in dice. This order, Della Porta says, far from being a particularity of gardens and plantations, "is to be found in each and every single thing in nature."[25] The figure of the quincunx is mentioned in other authors, both from classical antiquity and from the Renaissance; but the idea that this pattern transcends the realm of artificiality is original to Della Porta. Given that the Neapolitan polymath brings up the issue only to abandon it immediately after, the task of exploring it further was left in the hands of his readers. And it is Thomas Browne who undertakes it in a 1658 treatise on plantation techniques through the ages, titled *The Garden of Cyrus, or the Quincunciall Lozenge, or Net-*

work Plantations of the Ancients, Artificially, Naturally, Mystically Considered.

The Garden of Cyrus—published together with *Urn Burial,* one of the highest achievements in the history of English prose—begins as a survey of ancient techniques to design and plant gardens, but rapidly becomes a treatise on the quincunx. The garden is an ideal beginning for Browne because it represents a liminal space, halfway between nature and artifice. His attention is then directed to human-made artifacts where he also corroborates the pervasiveness of the quincuncial design. Browne's examples include statues, paintings, jewelry, board games, surgical instruments, weapons, battle formations, and the urban design of ancient cities. Browne insists particularly on anything that shows a reticular design, like windows, tapestries, textiles, and embroideries. Textile arts, referred to by Browne as *texturies,* are dominated by the quincuncial design. It is at this point that Browne suddenly switches terminology, discards the term "quincunx" and replaces it simply with "texture."[26] In the same way that the textile arts are essentially quincuncial, the quincunx is a textile model, a texture. Its presence in plants and animals shows, says Browne, that far from being a human artifice, it constitutes a natural model and, since nature is the work of God, a divine pattern.

The quincunx can be found on hard, sharp, or rough textures, on "rhomboidal protuberances" (like the skin of a pineapple), on soft and elegant membranes (like the spider web), or on harmonious and angular textures like those found in honeycombs.[27] As he moves from the vegetable to the animal kingdom, it becomes clear that the quincunx, this ubiquitous texture, is also the skin. Some of the most astonishing examples for Browne are the tail of the beaver, the skin of certain snakes, the bumpy skin of fowl, fish scales, and, of course, human epidermis: "The same is also observable in some parts of the skin of man, in habits of neat texture and therefore not unaptly compared unto a net."[28] So too, the insides of man are of quincuncial texture, "not only the first *subtegmen* or warp of his formation, but the netty fibres of the veins and vessels of life."[29] And this, says Browne paraphrasing Psalm 139, confirms the Scriptures when man addresses God and thanks Him: "Thou hast curiously embroidered me, thou has wrought me up after the finest way of texture, and as it were with a needle."[30] The terms "embroider" and "texture" are additions by Browne and perhaps shed light on that mysterious passage of *Religio Medici* where the young physician notices that whereas God created the rest of the world with "a blast of his mouth . . . at his bare word . . . , in the frame of man . . . he played the sensible operator, and seemed not so much to create as *make* him."[31] This "making" of man is the quincuncial intertwining of veins, weaving of tendons, braiding of muscles, and embroidery of the skin. And from the animal kingdom Browne ascends to the heavens because the alignment and disposition of planets is also quincuncial, or "textured"; so are the invisible and incorporeal hierarchies of angels. In fact, if we plant gardens and manufacture artifacts quincuncially we do so because that is how God designs. Browne, who likes to imagine God as a majestic and benevolent hand, conceives the task of creation as manual, and the quincunx, as the design that contrives the physiognomy of the universe. Ultimately, as the metaphysical texture of all that is, the quincunx functions as the skin of the real.

I jump back to the beginning, to the curious theory of a metempsychosis of opinions, ideas, images, motifs that runs through history via underground channels, emerging here and there to inhabit new bodies and minds. The quincunx too transmigrates, from Quintilian and Virgil to Della Porta and Browne, and from there to Coleridge's marginal notes on *The Garden of Cyrus*, to James Joyce's "Grace," to Borges's "Death and the Compass," to Charles Palliser's mystifying best-seller, and to the first chapter of Sebald's magnificent *The Rings of Saturn*. Its Dubliner transmigration is particularly interesting because it reveals yet another line of connection between the Renaissance and the early twentieth century. In "Grace"—written in 1905 and published in 1914—Joyce, who knew Browne's work well, employs the quincuncial order for the seating arrangements of five friends at church. They have attended a spiritual retreat for businessmen, eager to wash away their sins and calm their anxieties. The church is packed to the brim. Two sit in one row, one of them in the row behind them, and the other two behind him, the party settling down "in the form of a quincunx."[32] The reference to the archaic geometric disposition elevates us over the characters at the very moment that the priest begins a fatuous speech in which he compares grace and redemption with matters of accounting. Forming a quincunx, the five friends delineate a surface underneath which it is revealed, as though through tracing paper, the frivolous, epidermal reality of a place and a situation that are supposed to represent a refuge for the spirit.

The manifestation of the profound at surface level (or *as* surface) is one of the central themes of *Dubliners*, a collection of stories where nothing remains veiled, where everything comes afloat. Eveline's indifferent look when, at the last moment, she decides not to board the ship that would have taken her and her fiancé, a sailor named Frank, to Buenos Aires, reveals to him the harsh reality of her lack of affection. In "A Little Cloud," Little Chandler's fit of rage directed at his infant son expresses in a nutshell years of frustration and the profound disappointment with the life he has chosen. And in "The Dead," Gretta's unconditional devotion to her long-dead teenage sweetheart, washes up like the body of a drowned man, to show Gabriel, her husband, that there is no transcendent afterlife or great beyond, and that the living and the dead coexist overcrowded in a world that is brutally flat and homogenous. The last two pages of "The Dead," among the best pages written in the twentieth century, end with the image of snow falling all over Ireland. Once again, Joyce elevates the reader and we see his country as God would see it; only in Joyce's universe there is no transcendence. We see the snow fall over the central plains, over treeless hills, over peat swamps, over the ocean waves, over cemeteries, over old Celtic crosses, over all of Ireland blanketing at once under its snowy veil, the living and the dead. And scribbled on the endless white surface, in letters that shine like a raw body without skin, the news that God, as Stephen Dedalus would later put it, is but a shout in the street—the occasional surfacing of fear and shame, of pain and rage, to which modern secularism has reduced all that was left of the religious experience.

One last time I go back to Oxford, to that winter of 1629 or 1630, to the Bodleian courtyard, to the window of the medical school from which the tanned skin of a man hangs sadly, and to young Thomas Browne, who observes the spectacle from a low angle perspective. The letter where Browne, some four decades later, recalls this sight is entirely dedicated to the issue of skin. The study of skin was one of Browne's lifelong passions. After graduating from Oxford, the fledgling physician spent three years on the Continent. First, Montpellier and Padua, where he treated patients with bizarre skin afflictions and studied with pioneers of modern dermatology. Finally, Leiden, where he wrote his doctoral thesis on one of the most terrifying evils of his time: smallpox. In his *Observations in Anatomy* he acknowledges that the skin is the organ of touch, he exhorts students to study it closely, and he even recommends techniques to remove it efficiently.[33] But Browne's interest in flaying techniques was much more than just scientific curiosity; it was also an aesthetic and even a metaphysical concern. In one of his strangest texts, entitled *Musaeum Clausum*, Browne imagines the catalog of a nonexistent, or utopian, cabinet of curiosities. Apart from books and artifacts of all sorts, this collection holds a significant number of illustrations. Among them: some that depict "the exact method of flaying men alive, beginning between the shoulders," practiced by the Ottoman Turks according to Tomasso Minadoi in his *Wars between the Turks and the Persians*.[34] Minadoi's book, first published in English in 1595, contains grisly descriptions of said mode of execution but much to Browne's chagrin, no illustrations.

Browne was not the only one among his contemporaries who craved for images of flaying. Many anatomy manuals included illustrations of bodies without skin; there are even examples of frontispieces in which the author's name and title appear printed on human skin. The work of Spanish physician Juan de Valverde de Hamusco, *Historia de la composición del cuerpo humano* (1556; see fig. 4) features the shocking image of a man who has flayed himself. The body stands triumphantly holding his skin like a trophy, and it seems to be saying: "Here I am. This is me. *Ecce cutis!*" The engraving confirms both the importance given in the Renaissance to the discovery of the skin and an overall perception of the anatomical revolution as an exciting, even heroic enterprise. But the interest in such imagery exceeded the realm of anatomy. Skin also captivated the imagination of painters and sculptors in the Renaissance. Three famous flaying scenes from the classical and Christian traditions were the subject of an unprecedented number of visual renditions between the fifteenth and seventeenth centuries. First there's Sisamnes, a corrupt judge who, according to Herodotus, was flayed alive by order of Cambyses, the king of Persia. Perhaps the most famous version of the scene is that by Dutch painter Gerard David (1498; see fig. 5). In David's gory vision, Sisamnes lies on a table while five men remove the skin from his arms, legs, and chest in front of an imperturbable crowd. The similarity with the iconography from contemporary anatomy manuals is striking. David indeed presents the scene of torture as a medical spectacle.

The second example is the apostle Bartholomew, who was flayed alive for converting the brother of the King of Armenia to Christianity. The most famous pictorial rendition of Bar-

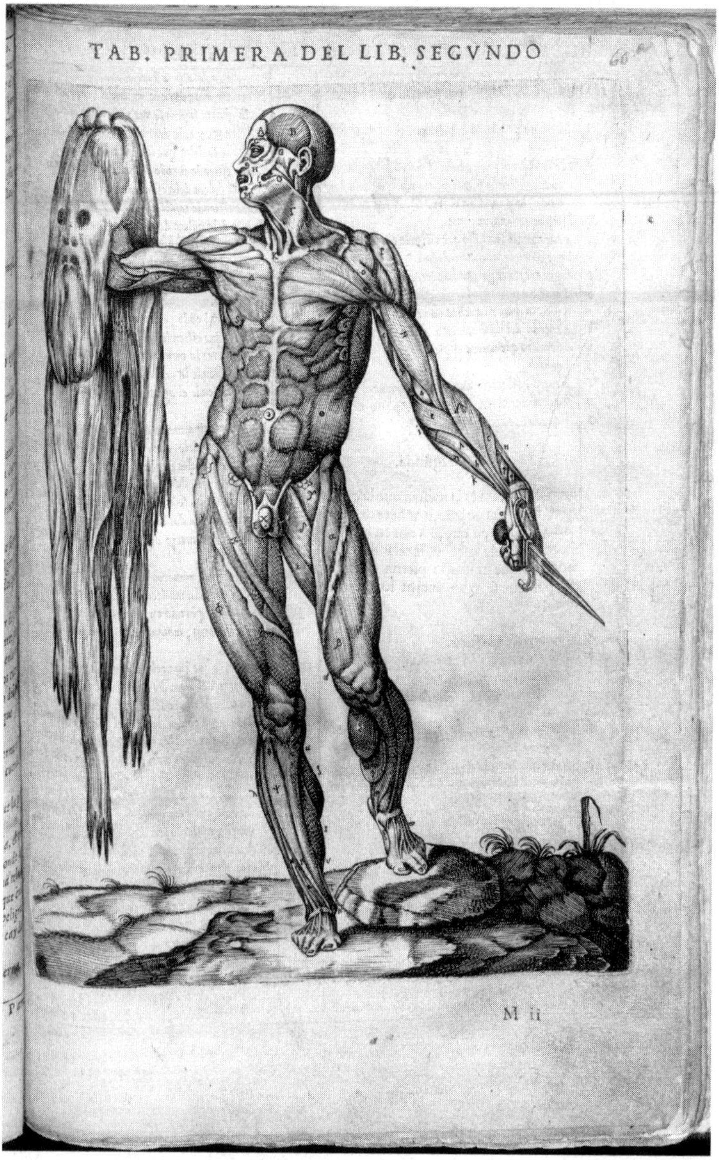

4. Juan de Valverde de Hamusco, *Historia de la composición del cuerpo humano* (Rome, 1556). Courtesy of the University of Chicago Library Special Collections Research Center.

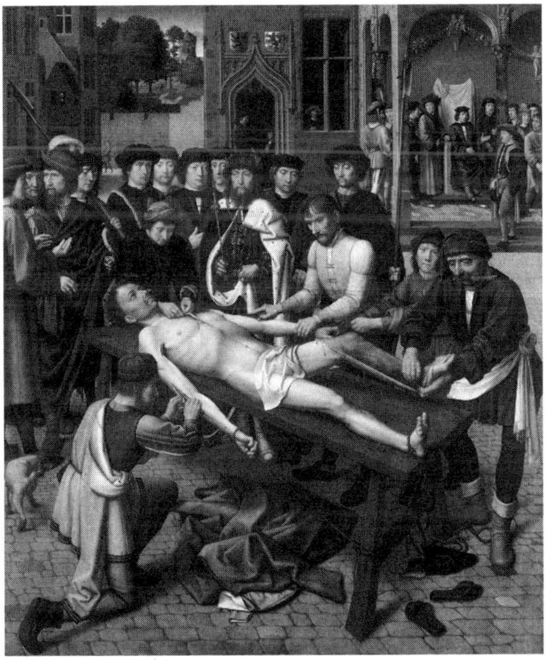

5.
Gerard David, *The Judgement of Cambyses* (1498).

tholomew is, as I mentioned before, Michelangelo's *Last Judgment*. There is a version of the scene, however, that depicts his flayed body as an anatomical spectacle. It is a sculpture by Marco d'Agrate, held at the Duomo of Milan, and dating from 1562 (see fig. 6). The artist presents the apostle with his skin wrapped around his body, like a toga. His tendons, muscles, and veins are carved in such vivid detail that after he saw it in the 1860s Mark Twain said: "I am very sorry I saw it, because I shall always see it now."[35]

And finally there is, of course, Marsyas. In book 6 of the *Metamorphoses* Ovid tells the story of the unfortunate satyr who dared challenge Apollo to a musical contest, lost, and was flayed alive by the god. The most famous rendition of the scene is Titian's, from the early 1570s, but they exist in an astonishing number, both in painting and in sculpture. Paolo Veronese, Luca Giordano, Giuseppe de Ribera, and others produced their

6. Marco D'Agrate, *Saint Bartholomew* (1562).

7. Melchior Meier, *Apollo and Marsyas and the Judgment of Midas* (1581).

own versions. One engraving in particular, by Melchior Meier (1581; see fig. 7), depicts the scene with perspicuous anatomical detail. In Meier's version, Marsyas has already been flayed. Apollo holds his skin in his left hand and a dagger in his right, like Michelangelo's Bartholomew. But while in Meier's engraving the hand of the god that holds the skin is pointing upward in a celebratory manner (Apollo was also the god of medicine) and the one with the dagger downward, in Michelangelo's fresco the knife, symbol of liberation from the body, points up, toward Christ, and the skin, representing the earthly carcass that we leave behind, points to the world.

In Edgar Wind's classical reading of the Renaissance's obsession with this myth, the scene is interpreted as a Platonic mystery that allegorizes ridding oneself of bodily garments to initiate the intellectual ascent toward the divine.[30] A closer and more literal reading of Ovid's passage in the light of the early modern anatomical fixation points in a very different exegetic direction:

> "Why do you tear me from myself?" he cried. "Oh, I repent! Oh, a flute is not worth such a price!" As [Marsyas] screams, his skin is stripped off from the surface of his body, and he is all one wound (*nisi vulnus erat*); blood flows down on every side, the sinews lie bare, his veins throb and quiver with no skin to cover them: you could count the entrails as they palpitate and the vitals showing clearly in his breast.[37]

The cries of the satyr are quickly muffled by the cold description of his vivisection. Marsyas, torn from himself, ceases to be Marsyas and becomes a wound, a raw, unidentifiable body, ready to be studied by the reader, who in turn becomes an eyewitness, encouraged to focus on the stripping of the skin and the naked veins, counting the throbbing organs and the glistening tissue, not unlike a student at an anatomy lesson. More than a Neoplatonic allegory for spiritual anabasis, this pictorial fascination with the flaying of Marsyas, in all its graphic and gory detail, seems to be yet another sign of the curiosity and excitement shared by artists, physicians, and intellectuals in general after having discovered the body and its long-neglected, multilayered, profoundly symbolic envelope: the skin.

In late April of 1912 James Joyce took a train from Trieste to Padua, the old city of the anatomists, where he was scheduled to sit for a series of exams that would certify him as an English teacher in Italy. After a long oral examination, the Irish expatriate was asked to write two essays, one in Italian and one in English, to demonstrate proficiency in both languages. For the essay in English, Joyce chose as his topic the centennial of Charles Dickens's birth. The second essay, written in a spotless Italian, discusses "modern materialism." Joyce was not referring here to materialism *strictu sensu*. He meant neither ancient atomism nor nineteenth-century Marxism, rather a certain sensualism and hedonism that has only loosely been associated, since classical antiquity, with the philosophies of Epicurus and Lucretius. In the essay, Joyce portrays his contemporaries as creatures solely committed to pleasure and lost to the frenzy of consumption; beings mobilized purely by impulses and stimuli, avidly seeking strong emotions, intense enjoyment, and little else. And he sees in the modern world—that world that he would years later present through the disenchanted eyes of Stephen Dedalus and Leopold Bloom—an inexhaustible bazaar of pleasures and delights, but also a wasteland of disillusion, tedium, humiliation, and shame. Modern man, Joyce argues in his Italian essay, has lost all sense of decorum vis-à-vis the quest for bodily bliss. The pursuit of happiness, made into a foundational human right by the American constitution and forming, along with the ideals of the French revolution, the very skeleton of the modern ethos, is nothing but a shameless and egotistical hunt for superficial enjoyment, Joyce suggests. Self-indulgent, narcissistic, and arrogant, modern man is pure surface, pure body, nothing but skin. The origins of this *Weltanschauung*, of this material man, Joyce locates in the early days of modernity. He then concludes: "One could say that the modern man has an epidermis rather than a soul."[38] The title of the essay is "The Universal Literary Influence of the Renaissance."

Joyce is not totally misguided when he associates this type of materialism, understood as sensualism, with the Renaissance. His connection is one of many lines cast from the early days of the twentieth century to the early days of modernity. These two epochs, remarkable in their thirst for knowledge and invention, are united by the shared conviction that there is no mystery more pressing and fascinating than that of the human. The first one discovers (or invents) the human body by digging its way through the dark, and hitherto unexplored, cavities of the flesh; the second one invents (or discovers) the unconscious and drags out layers of concealed interiority, our deep tectonic plates, leaving them to dry on the surface. The first one announces for the first time the organic and medical importance of the skin, while the second one consolidates dermatology as a vital branch of medicine. The first one rediscovers nudity and eroticism in the arts; the second one invents film, that moving image imprinted on a skin-like membrane. The first one sees an explosion of philology and the study of literary form; the second one witnesses the emergence of formalisms and avant-garde movements that anatomize language and push it to the very limits of its possibilities. But, above all, these two epochs cultivate a sophisticated and self-aware fondness for ambiguity and paradox; they vigorously resist their immediate cultural heritage; they are as irreverent about their fascination with the ephemeral and the banal as they are unrepentant in their ambition to dig deeper and deeper; and when we hit rock bottom, when we can't go any further, they both show us that even that is nothing but surface.

ACKNOWLEDGMENTS

My gratitude goes first to friends, readers, and editors in Buenos Aires, where the original version of this book was published in 2015. Special thanks go to my mother, Angela Signorini, whose careful and incisive comments were an invaluable help through the process of editing the manuscript. I also want to thank Eugenio Monjeau, María Zorraquín, Juan Zorraquín, and Damián Tabarovsky for their generous feedback, for their friendship, and for their support. At the University of North Carolina at Chapel Hill, I want to thank Jessica Wolfe, whose kind and intelligent guidance allowed me the scholarly freedom that ultimately opened the door for this book of essays. Also in North Carolina, I thank my friend Reid Barbour, as well as Eric Downing, Mary Floyd-Wilson, and Gregg Flaxman, for their constant support; all of them were also great sources of inspiration. I am no less grateful to many of my colleagues at the University of Chicago, who over the past four years read my work and encouraged me to pursue it. Among them, I acknowledge with special esteem Ada Palmer, Richard Strier, David Bevington, Miguel Martínez, William Mazzarella, Christopher Wild, Joshua Scodel, Caryn O'Connell, Benjamin Fong, Anna Seastrand, and David Simon. Other friends and readers of my work here in the United States are John Hamilton, David Damrosch, Ross Hamilton, Valeria Finucci, Jorge Silvetti, and Mariano Siskind; to them I extend my most heartfelt gratitude. On the other side of the pond, Michael Silk, Rosa Mucignat, Claire Preston, and Bashir Abu-Manneh were early supporters and enthusiasts of my work, and for that I thank them immensely. Last but certainly not least, I want to thank Karen Darling at the University of Chicago Press, who believed in this project from the beginning and whose support has been unwavering. This book is dedicated to my wife, my partner, and my best friend, Erin June Lodeesen, for her unconditional love and support, and for her inestimable help during the arduous process of transforming the Spanish text into an English one, but more importantly because whenever I think of touch, I first and foremost think of her.

NOTES

PREFACE

1. Johann Gottfried Herder, "Zum Sinn des Gefühls," in *Werke*, vol. 4 (Frankfurt-am-Main: Deutscher Klassiker Verlag, 1994): 233; "I feel! I am!" Translation my own.

2. Erich Auerbach, *Mimesis: The Representation of Reality in Western Literature*, trans. Willard R. Trask (Princeton, NJ: Princeton University Press, 2003): 3.

A SQUEEZE OF THE HAND

1. Luxorius, *Opera Omnia*, trans. Art Beck (Los Angeles: Otis Books, 2012): 158.

2. Some examples of notable publications from the last two decades are Gabriel Josipovici, *Touch* (New Haven, CT: Yale University Press, 1996); Yvette Hatwell, Arlette Streri, and Edouard Gentaz, *Touching for Knowing: Cognitive Psychology of Haptic Manual Perception* (Philadelphia: John Benjamins Publishing, 2000); Édouard Gentaz, *La main, le cerveau et le toucher* (Paris: Dunod, 2009); Constance Classen (ed.), *The Book of Touch* (Oxford: Berg, 2005); Constance Classen, *The Deepest Sense: A Cultural History of Touch* (Urbana-Champaign: University of Illinois Press, 2012); Mark Paterson, *The Senses of Touch: Haptics, Affects, and Technologies* (Oxford: Berg, 2007); Mark Paterson and M. Dodge (eds.), *Touching Place, Placing Touch* (Surrey: Ashgate, 2012); Fiona Macpherson (ed.), *The Senses: Classic and Contemporary Philosophical Perspectives* (Oxford: Oxford University Press, 2011); and Mathew Fulkerson, *The First Sense: A Philosophical Study of Human Touch* (Cambridge, MA: MIT Press, 2014).

3. Robert Jütte, *A History of the Senses: From Antiquity to Cyberspace*, trans. James Lynn (Cambridge: Cambridge University Press, 2005): 16.

4. Riegl introduces the term in 1902 as an amendment to the word *taktisch*, which he originally uses in his study of late Roman art industry. See Aloïs Riegl, *Das holländische Gruppenporträt*, vol. 1 of 2, ed. Karl M. Swoboda (Vienna: Österreichischen Staatsdruckerei, 1931): 275. See also Aloïs Riegl, *Historical Grammar of the Visual Arts*, trans. Jacqueline E. Young (New York: Zone Books, 2004): 187.

5. Aloïs Riegl, *The Group Portraiture of Holland*, trans. Evelyn M. Kain and David Britt (Los Angeles: Getty Research Institute Publications and Exhibitions Program, 1999): 259.

6. Gilles Deleuze and Félix Guattari, *A Thousand Plateaus: Capitalism and*

Schizophrenia, trans. Brian Massumi (Minneapolis: University of Minnesota Press, 2011): 492–500.

7. Aristotle, *On the Soul*, ed. W. S. Hett, Loeb Classical Library (Cambridge, MA: Harvard University Press, 1974): 126–35 (422b–424a).

8. Aristotle, *History of Animals* (494b). See also *Parts of Animals*, ed. A. L. Peck, Loeb Classical Library (Cambridge, MA: Harvard University Press, 1983): 370–71 (687a–b).

9. Plato, *The Symposium*, ed. Christopher Gill (London: Penguin, 1999): 7 (175d).

10. Plato, *Lysis-Symposium-Gorgias*, ed. W. R. M. Lamb, Loeb Classical Library (Cambridge, MA: Harvard University Press, 1961): 204 (211b).

11. Plotinus, *Ennead* 1.6.7, in *Enneads*, vol. 1, ed. A. H. Armstrong, Loeb Classical Library (Cambridge, MA: Harvard University Press, 1964): 253–55; and *Ennead* 4.6.2, in *Enneads*, vol. 4, ed. A. H. Armstrong, Loeb Classical Library (Cambridge, MA: Harvard University Press, 1984): 325–27.

12. Plotinus, *Ennead* 6.9.4, in *Enneads*, vol. 7, ed. A. H. Armstrong, Loeb Classical Library (Cambridge, MA: Harvard University Press, 1988): 315–19.

13. See Diogenes Laertius, *Life of Epicurus*, in *Lives of Eminent Philosophers*, vol. 2, ed. R. D. Hicks, Loeb Classical Library (Cambridge, MA: Harvard University Press, 1958): 557 (10.28); and Aristotle, *On Sense and the Sensible*, in *On the Soul, Parva Naturalia, On Breath*, ed. W. S. Hett, Loeb Classical Library (Cambridge, MA: Harvard University Press, 1957): 245–47 (442 a 29).

14. Lucretius, *On the Nature of Things*, ed. W. H. D. Rouse, Loeb Classical Library (Cambridge, MA: Harvard University Press, 2002): 26–27 (1.304).

15. Lucretius, *On the Nature of Things*, 38–39 (1.451–54).

16. Lucretius, *On the Nature of Things*, 128–28 (2.434–35).

17. See Lucretius, *On the Nature of Things*, 332–33 (4.722 ff.).

18. The German word for "concept," *Begriff*, comes from the verb *greifen*, which means "to hold," or "to grab."

19. Lorenzo Ghiberti, *I commentari* (Naples, 1947): 108. Translation my own.

20. See Riegl, *Historical Grammar*, 187.

21. Laura Marks, *Touch: Sensuous Theory and Multisensory Media* (Minneapolis: University of Minnesota Press, 2002): ix.

22. Herman Melville, *Moby Dick, or The Whale* (New York: Modern Library, 1950): 3.

23. Melville, *Moby Dick*, 414.

24. Melville, *Moby Dick*.

25. Melville, *Moby Dick*, 414–15.

26. Melville, *Moby Dick*, 415.

27. *The Letters of Herman Melville*, ed. Merrell R. Davis and William H. Gilman (New Haven, CT: Yale University Press, 1960): 108.

SIX FINGERS

1. Homer, *The Iliad*, trans. Robert Fagles (London: Penguin, 1990): 454 (16.366–69).

2. Homer, *Iliad*, 455 (16.384–401).

3. Herodotus, *The Histories*, trans. Robin Waterfield (Oxford: Oxford University Press, 1998): 324–25.

4. Homer, *Iliad*, 490 (19.64).

5. Homer, *Iliad*, 207 (6.357–58).

6. Homer, *Iliad*, 483–87 (18.481–608).

7. See Robert Graves, *The Greek Myths*, vol. 1 (Edinburgh: Penguin, 1955): 185–88; and Karl Kerényi, *The Gods of the Greeks*, trans. Norman Cameron (New York: Grove Press, 1960): 83–88.

8. See Aristotle, *Parts of Animals*, 370–71 (687a–b).

9. See also Pablo Maurette, "The Organ of Organs: Vesalius and the Wonders of the Human Hand," *Journal of Medieval and Early Modern Studies* 48.1 (2018): 1–20.

10. Erich Auerbach, *Mimesis: The Representation of Reality in Western Literature*, trans. Willard R. Trask (Princeton, NJ: Princeton University Press, 2003): 3.

11. Homer, *The Odyssey*, trans. Robert Fagles (New York: Penguin, 1996): 402 (19.429–32).

12. Homer, *Odyssey*, 405 (19.528–39).

13. Genesis 27:21–22, trans. Robert Alter (London: W. W. Norton, 1996): 139–40.

14. Genesis 27:23–24.

15. Homer, *Odyssey*, 265 (11.248–49).

16. Homer, *Iliad*, 562–63 (23.99–104).

17. See Shirley Darcus Sullivan, *Psychological Activity in Homer: A Study of Phren* (Don Mills: Carleton University Press, 1988): 25.

18. Homer, *Iliad* 607 (24.582–86).

TORN TO PIECES

1. See Melissa Macauley, *Social Power and Legal Culture: Litigation Masters in Late Imperial China* (Stanford, CA: Stanford University Press, 1998). For more on *lingchi*, see *Death by a Thousand Cuts*, by Timothy Brook, Jérôme Bourgon, and Gregory Blue (Cambridge, MA: Harvard University Press, 2008).

2. See Pablo Maurette, "*De rerum textura*: Lucretius, Fracastoro, and the Sense of Touch," *Sixteenth-Century Journal* 45.2 (2014): 309–30.

3. See Giambattista Bussini, *Lettere a Benedetto Varchi sopra l'assedio di Firenze* (Florence, 1860): 241.

4. Lucretius, *On the Nature of Things*, 94 (2.1–3).

5. Five crucial books for understanding the importance of the rediscovery of Lucretius in Renaissance Europe are Valentina Prosperi's *Di soavi licor gli orli del vaso: La fortuna di Lucrezio dall'Umanesimo alla Controriforma* (Turin: N. Aragno, 2004), Lisa Piazzi's *Lucrezio: Il "De rerum natura" e la cultura occidentale* (Naples: Liguori Editore, 2009), Stephen Greenblatt's *The Swerve: How the World Became Modern* (New York: W. W. Norton, 2011), Gerard Passannante's *The Lucretian Renaissance: Philology and the Afterlife of Tradition* (Chicago: University of Chicago Press, 2011), and Ada Palmer's *Reading Lucretius in the Renaissance* (Cambridge, MA: Harvard University Press, 2014).

6. Giovanni Pontano, *I dialoghi* 238–39. Cited and translated by Yasmin Haskell, "Religion and Enlightenment in the Neo-Latin Reception of Lucretius," in *The*

Cambridge Companion of Lucretius, edited by Stuart Gillespie and Philip Hardie (Cambridge: Cambridge University Press, 2007): 189.

7. Marsilio Ficino, *Letters* 11.25 and *Platonic Theology* 14.10.

8. Lucretius, *On the Nature of Things*, 200 (3.161–67).

9. Lucretius, *On the Nature of Things*, 128 (2.434–35).

10. Erwin Schrödinger, *Nature and the Greeks and Science and Humanism* (Cambridge: Cambridge University Press, 1996): 88.

11. Schrödinger, *Nature and the Greeks*, 88.

12. Lucretius, *On the Nature of Things*, 23 (1.250–64): "Postremo pereunt imbres, ubi eos pater aether / in gremium matris terrain praecipitavit; / arboribus, crescent ipsae fetuque gravantur; / hinc alitur porro nostrum genus atque ferarum; / hinc laetas urbes pueris florere videmus / frondisferasque novis avibus canere undique silvas; / hinc fessae pecudes pingui per pabula laeta / corpora deponent, et candens lacteus umor / uberibus manat distentis; hinc nova proles / artubus infirmis teneras lasciva per herbas / ludit lacte mero mentes perculsa novellas. / haud igitur penitus pereunt quaecumque videntur, / quando alid ex alio reficit natura, nec ullam / rem gigni patitur nisi morte adiuta aliena."

13. Cited by Deborah Hayden in *Pox: Genius, Madness, and the Mysteries of Syphilis* (New York: Basic Books, 2003): 14.

14. Girolamo Fracastoro, *Syphilis, or the French Disease*, in *Latin Poetry*, trans. James Gardner (Cambridge, MA: Harvard University Press, 2013): 22–25 (1.331–64).

15. Gareth Williams, *Angel of Death: The Story of Smallpox* (London: Palgrave Macmillan, 2010): 34. Recent studies on the DNA of the mummified body of a Lithuanian child who died in 1654 show that the *variola* virus might not be as old as historians of medicine used to think; in fact, the study speculates that smallpox as we know it might be an early modern novelty (see "17th Century Variola Virus Reveals the Recent History of Smallpox," *Current Biology*, online edition, December 8, 2016).

16. Lucretius, *On the Nature of Things*, 277 (4.5–9).

17. Lucretius, *On the Nature of Things*, 265 (3.972–75).

18. Lucretius, *On the Nature of Things*, 29 (1.328).

19. Lucretius, *On the Nature of Things*, 363 (4.1107–14).

20. See Lucretius, *On the Nature of Things*, 121 (2.320–30).

21. From the documentary *Joel-Peter Witkin: An Objective Eye* (dir. Thomas A. Marino, USA, 2013).

ELEMENTS OF PHILEMATOLOGY

1. "I celebrate the affable games / that some despise in vain, / the joys of sensible souls, / the sweet mysteries of the kiss." Claude-Joseph Dorat, *Les baisers* (Paris: Lambert, 1770): 101. Translation my own.

2. Cited in *L'art du baiser: Les plus beaux baisers de la literature* (Paris: Gallimard, 2011): 15. Translation my own.

3. Briffault notes that in ancient Egypt the word used to refer to the kiss also meant to eat. See Robert Briffault, *The Mothers: A Study of the Origins of Sentiments and Institutions*, 3 vols. (New York: Macmillan, 1927): 1:120.

4. Søren Kierkegaard, *The Seducer's Diary*, trans. Edward V. Hong and Edna H. Hong (Princeton, NJ: Princeton University Press): 160.

5. Kierkegaard, *Seducer's Diary*, 161.

6. Kierkegaard, *Seducer's Diary*, 161.

7. To be fair to Kierkegaard, *Delfino* was never published in Patrizi's lifetime. Paul Oskar Kristeller discovered the text in the mid-twentieth century, and the very first edition appeared in 1975. One of the earliest historical studies of the different types of kisses also comes from the early modern period. Martin von Kempe's *Opus polyhistoricum . . . de osculis* (Frankfurt, 1680) is a gargantuan volume of over a thousand pages that surveys at least twenty different kinds of kisses. For more on this, see *The Kiss in History*, ed. Karen Harvey (Manchester: Manchester University Press, 2005): 187.

8. Francesco Patrizi, *Delfino ovvero del bacio*, in *Lettere ed opuscoli inediti*, ed. D. Aguzzi Barbagli (Florence: Istituto Nazionale di Studi sul Rinascimento, 1975): 142. Translations of Patrizi are my own.

9. Patrizi, *Delfino*, 144.

10. *The Kisses of Johannes Secundus: Basia & Epithalamium*, trans. F. X. Matthews (Kingston, RI: Winecellar Press, 1984).

11. For the Latin text: Jean Second, *Les Baisers*, ed. Olivier Sers (Paris: Les Belles Lettres, 2006): 46.

12. *Basium* 4.8–9; *Basium* 10.10–14.

13. A distich attributed to Plato by Diogenes Laertius might very well be one of the oldest kiss poems that we know of: "My soul was on my lips as I was kissing Agathon. Poor soul! She came hoping to cross over to him." Cited by Nicholas Perella in *The Kiss Sacred and Profane* (Berkeley: University of California Press, 1969): 7.

14. Maffio Venier, *Canzoni e soneti* (Venice: Corbo e Fiore Editori, 1993): 258. Translation my own.

15. See Pablo Maurette, "Shakespeare's *Venus and Adonis* and Sixteenth-Century Kiss Poetry," *English Literary Renaissance* 47.3 (2017): 1–25.

16. Sir Philip Sidney, *Astrophil and Stella*, in *Selected Writings*, ed. Richard Dutton (Manchester: Flyfield Books, 1987): 70.

17. Sydney, *Astrophil and Stella*, 70.

18. Guy de Maupassant, "The Kiss," in *The Complete Short Stories*, trans. Artine Artinian (New York: Hanover House, 1955): 938.

19. Maupassant, "The Kiss," 938.

20. Maupassant, "The Kiss," 939.

21. Karl Ove Knausgaard, *Boyhood Island: My Struggle 3*, trans. Don Bartlett (London: Harvill Seeker, 2014): 438–39.

22. Julio Cortázar, *Hopscotch*, trans. Gregory Rabassa (New York: Pantheon, 1966): 33.

23. The *dialogo d'amore* ("love dialogue") is a subgenre of the philosophical dialogue, inspired in Plato's *Symposium*, that deals exclusively with issues pertaining love, and that proliferated between the late fifteenth and all through the sixteenth centuries. For more on this see Pablo Maurette, "Plato's Hermaphrodite

and a Vindication of the Sense of Touch in the Sixteenth Century," *Renaissance Quarterly* 68.3 (2015): 872–98.

24. Cited by Hugh Maclean (ed.), *Ben Jonson and the Cavalier Poets* (London: W. W. Norton, 1974): 134.

25. *L'impresa* only exists in manuscript form and it has never been published. For more on it, see Armando Maggi, "On Kissing and Sighing: Renaissance Homoerotic Love from Ficino's *De amore* and *Sopra lo amore* to Cesare Trevisani's *L'impresa* (1569)," in *Journal of Homosexuality* 49.3–4 (2005): 331–32.

26. Antonio Rocco, *L'Alcibiade fanciullo a scola* (Rome: Salerno Editrice, 1988): 48. Translation my own.

27. Rocco, *L'Alcibiade*, 48.

28. Wilhelm Wundt, *Völkerpsychologie: Eine Untersuchung der Entwicklungsgesetze von Sprache, Mythus und Sitte*, 4 vols. (Leipzig: Alfred Kröner Verlag, 1920): 3:135. Cited by Perella, *The Kiss Sacred and Profane*, 6.

29. See Adam Phillips, "Plotting for Kisses," in *On Kissing, Tickling, and Being Bored: Psychoanalytic Essays on the Unexamined Life* (Cambridge, MA: Harvard University Press, 1993): 93–100. Phillips argues that the kiss is a type of eating that dispenses with food.

30. Lucretius, *On the Nature of Things*, 361–63 (4.1080–1110).

THE FRENCH CONNECTION

1. Cited in Jean-Claude Carrière and Umberto Eco, *This is Not the End of the Book*, a conversation curated by Jean-Philippe de Tonnac, trans. Polly McLean (London: Harvill Secker, 2011): 32.

2. Three of the main pioneers of tactile writing were French. The first one was Valentin Haüy, founder in 1784 of the first school for the blind, and dedicated reader of Diderot. The second, Charles Barbier, a captain during the Napoleonic Wars, introduced a system known as "night writing," or sonography, in 1808. This was crucial for the development of Braille, whose inventor, Louis Braille, presented it to the world in 1829 and it was officially adopted in 1854. Also in the nineteenth century, French physicians working at the St. Louis Hospital in Paris revolutionized the study of dermatology.

3. In a late essay, Edmund Husserl says that "the human being who knows himself to be in the world with living body and soul, who moves in space and works with his hands, as a manual worker, or in some other way with his living body, who in battle also fights with his living body is natural, always coconscious of his living body, acting through it upon his external world or experiencing through it a touch, a push, a wound." Edmund Husserl, "The Attitude of Natural Science and the Attitude of Humanistic Science: Naturalism, Dualism, and Psychophysical Psychology," in *The Crisis of European Sciences and Transcendental Phenomenology*, trans. David Carr (Evanston, IL: Northwestern University Press, 1970): 322.

4. Chrétien de Troyes, *The Knight of the Cart (Lancelot)*, in *Arthurian Romances*, trans. William W. Kibler (London: Penguin, 1991): 263.

5. Chrétien de Troyes, *Knight of the Cart*, 259.

6. Michel Henry, *Incarnation: A Philosophy of the Flesh*, trans. Karl Hefty (Evanston, IL: Northwestern University Press, 2015): 137.

7. Henry, *Incarnation*, 149.

8. Henry, *Incarnation*, 84–85.

9. Henry, *Incarnation*, 101. Not surprisingly, French Renaissance poets explored and exploited the motif of the kiss with particular enthusiasm. Pierre de Ronsard, Joachim du Bellay, Rémy Belleau, Jean-Antoine de Baïf, and many others between the sixteenth and seventeenth centuries imitated the Latin classics and Johannes Secundus, and even included *basia* in their poetic miscellanies.

10. Lucretius, *On the Nature of Things*, 29 (1.328).

11. For more on this see Ada Palmer's brilliant *Reading Lucretius in the Renaissance* (Cambridge, MA: Harvard University Press, 2014): 28.

12. See Michael Screech, *Montaigne's Annotated Copy of Lucretius: A Transcription and Study of the Manuscript, Notes, and Penmarks* (Geneva: Droz, 1998).

13. Bernard Rochot, *Les travaux de Gassendi sur Epicure et sur l'atomisme: 1619–1658* (Paris: Vrin, 1944): 30.

14. See Pierre Gassendi, *Institutio Logica et Philosophia Epicuri Syntagma* (London, 1660): 105.

15. See Gassendi, *Syntagma*, 122.

16. Cyrano de Bergerac, *Voyages to the Moon and the Sun*, trans. Richard Aldington (New York: Orion Press, 1962): 54.

17. Cyrano, *Voyages*, 80–81.

18. Cyrano, *Voyages*, 61.

19. Cyrano, *Voyages*, 115–16.

20. See pages 81–82.

21. Cyrano, *Voyages*, 77.

22. François de La Mothe Le Vayer, *Of a Man Born Blind*, trans. Kate Tunstall, in *Blindness and the Enlightenment: An Essay* (New York: Continuum, 2011): 232–33.

23. La Mothe Le Vayer, *Of a Man Born Blind*, 235–36.

24. According to Ernst Cassirer, the Molyneux problem was "the central question of eighteenth century epistemology and psychology." Cited in Mark Paterson, *Seeing with the Hands: Blindness, Vision, and Touch after Descartes* (Edinburgh: Edinburgh University Press, 2016): 33.

25. Denis Diderot, *Letter on the Blind for the Use of Those Who Can See*, trans. Kate Tunstall, in *Blindness and the Enlightenment: An Essay* (New York: Continuum, 2011): 173.

26. Diderot, *Letter on the Blind*, 176.

27. Diderot, *Letter on the Blind*, 177.

28. Diderot, *Letter on the Blind*, 198.

29. Diderot, *Letter on the Blind*, 199.

30. Diderot, *Letter on the Blind*, 197.

31. Plotinus, *Ennead* 1.6.7, in *Enneads*, vol. 1, ed. A. H. Armstrong, Loeb Classical Library (Cambridge, MA: Harvard University Press, 1964): 253–55.

32. See Henry, *Incarnation*, 113–16. Henry refers especially to Maurice Merleau-Ponty's *Le visible et l'invisible* (Paris: Gallimard, 1964): 302–4.

33. Henry, *Incarnation*, 136–44.

34. Henry's critique of Merleau-Ponty is not entirely fair. In the first part of *Phenomenology of Perception*, which is dedicated to the body, Merleau-Ponty does engage systematically with corporeality in and of its own, addressing subjects such as motor skills, proprioception, phantom limbs, kinesthesia, and affectivity.

35. Henry, *Incarnation*, 11–12.

36. Cited by Nicholas James Perella in *The Kiss Sacred and Profane: An Interpretive History of Kiss Symbolism and Related Religio-Erotic Themes* (Berkeley: University of California Press, 1969): 3.

SKIN DEEP

1. Friedrich Nietzsche, "Schopenhauer as Educator," in *Untimely Meditations*, trans. R. J. Hollingdale (Cambridge: Cambridge University Press, 1983): 129.

2. Sir Thomas Browne, *Religio Medici*, in *The Major Works*, ed. C. A. Patrides (London: Penguin, 1977): 66.

3. Browne, *Religio Medici*, 67.

4. Hugo von Hofmannstahl, *The Lord Chandos Letter and Other Writings*, trans. Joel Rotenberg (New York: New York Review Books, 2005): 121.

5. Hofmannstahl, *Lord Chandos Letter*, 122.

6. Hugo von Hofmannstahl, *Buch der Freunde* (Leipzig: In-Insel Verlag, 1922): 56. Translation my own.

7. Didier Anzieu, *The Skin Ego*, trans. Chris Turner (New Haven, CT: Yale University Press, 1989): 40.

8. *The Letters of Sir Thomas Browne*, ed. Geoffrey Keynes (London: Faber & Faber, 1931): 82.

9. See Claudia Benthien, *Skin: On the Cultural Border between Self and the World*, trans. Thomas Dunlap (New York: Columbia University Press, 2002): 53; and also François Dagognet, *La peau découverte* (Paris: Synthélabo, 1993).

10. See Martin Clayton and Ron Philo, *Leonardo Da Vinci: Anatomist* (London: Royal Collection Trust, 2012). The exhibit, *Leonardo Da Vinci: The Mechanics of Man*, took place between August and November of 2013 at the Queen's Gallery, in the Palace of Hollyroodhouse.

11. Paul Valéry, *Leonardo, Poe, Mallarmé*, Bollingen Series XLV.8, trans. Malcolm Cowley and James R. Lawler, in *The Collected Works of Paul Valéry*, ed. Jackson Mathews (Princeton, NJ: Princeton University Press, 1972): 82.

12. Paul Valéry, *Idée Fixe*, Bollingen Series XLV.5, trans. David Paul, in *The Collected Works of Paul Valéry*, ed. Jackson Mathews (Princeton, NJ: Princeton University Press, 1965): 32–333.

13. Filippo Tommaso Marinetti, *Il tattilismo*, in *Marinetti e il futurismo: Una antologia*, ed. Luciano de Maria (Verona: Mondadori, 1973): 244–45.

14. Marinetti, *Il tattilismo*, 247–50.

15. Giambattista Della Porta, *Of Human Physiognomy* (1585), *Phytophysiog-*

nomy (1588), *Celestial Physiognomy* (1603), and *Cheirophysiognomy* (published posthumously in 1677).

16. See Mark M. Anderson, *Kafka's Clothes: Ornament and Aestheticism in the Habsburg Fin de Siècle* (Oxford: Clarendon Press, 1992): 146.

17. *The Diaries of Franz Kafka 1910–1913*, ed. Max Brod and trans. Joseph Kresh (New York: Schocken, 1948): 92.

18. *Diaries of Franz Kafka 1910–1913*, 108.

19. *Diaries of Franz Kafka 1910–1913*, 179.

20. Cited in "Kafka Begins," by James Rolleston, in *A Companion to the Works of Kafka*, ed. James Rolleston (Rochester: Camden House, 2003): 14.

21. *Diaries of Franz Kafka 1910–1913*, 254.

22. Franz Kafka, *The Penal Colony: Stories and Short Pieces*, trans. Willa and Edwin Muir (New York: Schocken, 1961): 191.

23. Thomas Browne, *Religio Medici*, 135.

24. Browne, *Religio Medici*, 136.

25. Giambattista Della Porta, *Villae* (Frankfurt am Main, 1592): 197. Translation my own.

26. Thomas Browne, *The Garden of Cyrus*, in *The Major Works*, 343 ff.

27. Browne, *Garden of Cyrus*, 344–46.

28. Browne, *Garden of Cyrus*, 357.

29. Browne, *Garden of Cyrus*, 358.

30. The King James Version of this passage of the psalm simply reads: "My substance was not hid from thee, when I was made in secret and curiously wrought in the lowest parts of the earth" (Psalm 139.15).

31. Browne, *Religio Medici*, 105.

32. James Joyce, *Dubliners*, ed. Margot Norris (London: Norton & Norton, 2006): 149.

33. *Observations in Anatomy*, in *The Miscellaneous Writings of Sir Thomas Browne*, ed. Geoffrey Keynes (London: Faber & Faber, 1931): 303–5.

34. Thomas Browne, *Musaeum Clausum*, in *The Miscellaneous Writings*, 137.

35. Mark Twain, *The Innocents Abroad* (New York: New American Library, 1966): 126.

36. See Edgar Wind, *Pagan Mysteries in the Renaissance* (New Haven, CT: Yale University Press, 1958): 143–44.

37. Ovid, *Metamorphoses*, book 6 (385–91), trans. Frank Justus Miller (Cambridge, MA: Harvard University Press, 1977): 315.

38. *James Joyce in Padua*, ed. Louis Berrone (New York: Random House, 1977): 15.

INDEX